NATURAL DESIGN
The search for comfort and efficiency.

1. Ergonomics

Is there such a thing as *natural* design? Why is your chair uncomfortable? Do you get eyestrain? How do you switch this machine off?

Is it possible to make environments, tools and machines which complement human capabilities, or is man always going to be at odds with the things he makes?

Ergonomics is the study of these questions.

Chambers' Dictionary of Science and Technology defines ergonomics as "the study of work in relation to the environment in which it is performed and the personnel who perform it." Broadly speaking, there are two types of work. The first involves the handling of power, the second involves the handling of information.

Ergonomics is an interdisciplinary science, and like any other science it is based on observations and measurements. Its purpose is to quantify the relationship of the human body with the tools and the machines which men use, so as to make both people more comfortable and their artefacts more efficient. Increasingly, ergonomics is coming to mean the study of the structure, quality and value of the work itself. Ergonomics expects that machines and systems should be shaped by the needs of their users, rather than the other way around.

Ergonomics is the study of work and the working environment, but there are two types of work: power processing and information processing. In their book, *Human Factors*, Kantowitz and Sorkin divide work into four areas. 1·Human input into simple mechanisms. 2·Humans using more sophisticated tools and machines. 3·Humans controlling very sophisticated mechanisms which throw up data and 4·Humans as the managers of ultra-complex systems.

Although ergonomics is fundamental to the design process, its study is in its infancy. It is the latest stage in the development of industrial design as a technical discipline. Its origins are with the work study engineers early in the century. The first generation of industrial designers in the United States began to understand that people worked better in certain environments and that machines were more pleasant to use if sharp edges were given a gentle radius.

Work study engineers, who pioneered ergonomics, monitored time-and-motion in the factory environment.

4

But proper ergonomic research only began during the Second World War, when military scientists tried to improve the pilot's environment when they discovered that significant numbers of accidents were caused because aviators could not master the increasingly complicated instruments and controls in their cockpits.

During the Second World War badly designed altimeters were frequently the cause of aircraft accidents.

Although ergonomists are concerned with every aspect of man's relationship with machines, in practice their work is restricted to certain key areas. For instance: the design of information, handles, controls, the sitting position as well as visual, acoustic and thermal characteristics of the whole environment.

Although this sounds arid, ergonomics affects the character and performance of everything that is made, from books through to supersonic jets, kitchen machines and hi-fi. A major problem in ergonomic research is to understand the limits of human capability so as to provide a basis for the design of things which can exploit those capabilities to best advantage … in every sense.

As the marketplace becomes more fiercely contested, sound ergonomics will play an increasingly important part in the design of successful consumer products. 'Usability' is a selling point: educated consumers are not ashamed to complain when they cannot understand or operate new appliances.

6

This exhibition has been organised to introduce some of the fundamental concepts in ergonomics and to provide a demonstration of some of the ergonomic problems which face designers. It was conceived to establish the idea – by no means widely acknowledged – that the best products are those which respect the user, whether he is the installer of the original equipment, the owner of it, the service engineer or the repair man.

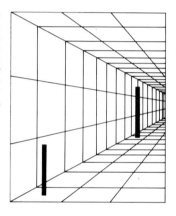

The corridor illusion is one of the most familiar exercises in perceptual psychology: the two vertical bars are, despite appearances, the same length. The fact that very simple data can mislead the brain is a problem that designers too rarely recognise.

2. PERCEPTION

The eyes absorb a disproportionate amount of our available energy.

It is not surprising that the problems of perception are perhaps the most significant part of the ergonomist's work.

Perception can be considered in two parts: *legibility,* which is to say "can we actually see such-and-such?" and *semantics,* which is to say "what does such-and-such mean?"

Legibility means discrimination and identification of information. It can be measured scientifically, but semantics is more abstruse because it is socially and culturally conditioned.

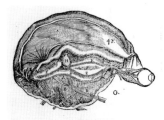

The eye and the brain, from Rene Descartes' *Traite de l'homme* (1664).

8

We will look at three very different things to help us understand some of the problems of perception: the book, the pilot's environment and the portable radio.

THE BOOK

The book is the most ancient data storage system, letters are the most ancient 'bits' of information. Long before the age of modern information technology, the literary critic, I. A. Richards, described a book as a "machine to think with".

The Mainz Psalter of 1457 was among the first books to be printed with moveable type.

fup me os fuu: dirertit eugt eugt viderut onli noftri   idifh dñe ne fileas : dñe ne difredas a me   r urge dñe z intedr nudino meo: drus me⁹ z dñs meus in tanfam meã   udita me fedm nfhnã miã dñe de⁹ me⁹: et no fupgaudeãr michi   on dират in cordibus fuis eugt eugt air nfe: nec dicant denorabim⁹ eũ,   tubefrãr z reuere añur funt: q̃ granilant̃ mahis mtis   n duant̃ rõfufione et reuerentia; q̃ maligna loquũt̃ fup me,   gulet̃ z letent̃ q̃ volũt nfhnã meã: et dirãt fenp mangficet̃ dñs qu̅ volunt̃ рас̃ fui rus   t lingua mea medirabit̃ nfhuã niã: tota die laudẽ niã, Ohre miufhus ut dlinquat in fe · [⁹ membo: nõ eft timor dei ante orfos rus   in dolofe egit in rofpeciu rus: ut inieniat iniquitas rus ad odiũ,   erba oris rus iniquitas et dolus: noluit intelligere ut bene agerer   niquitatẽ meditat⁹

PREFACE

A book is a machine to think with, but it need not, therefore, usurp the functions either of the bellows or the locomotive. This book might better be compared to a loom on which it is supposed to re-weave some ravelled parts of our civilisation. What is most important about it, the interconnection of its several points of view, might have been exhibited, though not with equal clarity, in a pamphlet or in a two-volume work. Few of the separate items are original. One does not expect novel cards when playing so traditional a game; it is the hand which matters. I have chosen to present it here on the smallest scale which would allow me to fit together the various positions adopted into a whole of some firmness. The elaborations and expansions which suggest themselves have been constantly cut short at the point at which I thought that the reader would be able to see for himself how they would continue. The danger of this procedure, which otherwise has great advantages both for him and for me, is that the different parts of a connected account such as this mutually illumine one another. The writer, who has, or should have, the whole position in his mind throughout, may overlook sources of obscurity for the reader, due to the serial form of the exposition. This I have endeavoured to prevent by means of numerous cross-references, forwards and backwards.

But some further explanation of the structure of the book is due to the reader. At sundry points - notably in Chapters Six, Seven, and Eleven to Fifteen - its progress appears to be interrupted by lengthy excursions into theory of value, or into general psychology. These I would have omitted if it had seemed in any way possible to develop the argument of the rest strongly and clearly in their absence. Criticism, as I understand it, is the endeavour to discriminate between experiences and to evaluate them. We cannot do this without some understanding of

In his *Principles of Literary Criticism* (1924) I.A. Richards described the book as a machine to think with, a conceit later echoed by the architect, Le Corbusier, in his remark about the nature of the house.

The Canadian academic, Marshall McLuhan (1911-1981) predicted the end of the written word as a means of communicating ideas. Paradoxically, his medium was ...the book!

In English words are made up out of twenty six visible signals, the letters of the alphabet.

Is there anything to choose between the legibility of different letter forms?

Not every letter form is as legible as the next. Different types have different characteristics; ergonomists and experimental psychologists try to ascertain which can be most easily read.

legibility

legibility

legibility

LEGIBILITY

legibility

The act of reading depends on the identification of whole words, not individual letters. In European languages there are two broad families of letter forms:

It is generally accepted that, while sans serif is effective for display work, traditional Roman type is easier to read in continuous text. The little feet, or 'serifs', lead the eye easily from one character to the next.

Roman and sans serif

The traditions of using quills, and later metal nibs, which both give the writer an opportunity to make broad upstrokes, or narrow downstrokes, reinforced the traditional virtues of patrician Roman letters resting on little 'feet' known as *serifs*.

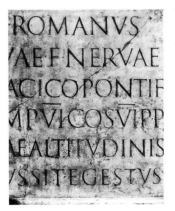

The inscription on Trajan's column (AD 113) is a superb example of Roman typography.

During the industrial revolution square cut letter forms were evolved for commercial use. These letter forms became known as *sans serif* and had the benefit of high visibility when used in display work.

Sans serif type became a sort of trademark of modernism during the early twentieth century. Avant-garde books were frequently set in sans serif type to demonstrate 'functionalism'.

Sans serif letter forms came to prominence in the nineteenth century where they were often used in coarse commercial applications.

Herbert Read's *Art Now* (1933) was set in sans serif to demonstrate that it was *avant-garde*.

Scientific and technical journals have traditionally been set in sans serif to demonstrate their essential practicality.

With sans serif type the limit for length is much shorter.

Although Burt discovered this scientifically, it is interesting to note that artisan printers had known it for centuries.

Yet Cyril Burt was able to demonstrate what printers already knew: that this sort of functionalism is only partially valid. While sans serif type is splendidly visible, it can most comfortably be read only in short lines and is most effective in capitals.

ngth for comfortable reading if a book is set in Roman type, as you can see from this sentence.

13

Graphic designers have to balance a number of variables in creating effective typography.

Which is the easiest to read –
upper or lower case?
monoline or thick & thin?
Roman or sans serif?
upright or inclined?
light or heavy?

UPPER or lower case?

monoline or thick and thin?

Roman or **sans serif?**

upright or *inclined?*

light or **heavy?**

There are no simple answers. Since the end of any general tradition of calligraphy, graphic designers and typographers have had to rediscover traditional principles. They are now more generally aware that, while old type-faces emphasised the differences between letter-forms and were therefore easier to read, sans serif typefaces tend to accentuate the similarities.

This booklet has been set in Caslon, a type designed by a Worcestershire gun engraver in the eighteenth century.

The American Declaration of Independence (1776) was set in Caslon.

THE PILOT'S ENVIRONMENT

Flight takes man's triumph over Nature to the limit and the pilot's environment is – quite literally – a cockpit where the human frame has to perform at the limit of its capabilities. The pilot has to make judgements based on observation of his environment and on information gleaned from his instruments, very often with no time or opportunity empirically to check if the data is correct.

Because so much depends on a pilot's performance, a great deal of the leading ergonomic research has taken place in the aerospace industries. The Institute of Aviation Medicine at Farnborough's Royal Aircraft Establishment is one of the world's great ergonomic study centres.

Aircraft designers and ergonomists need to understand how pilots see and understand their instruments. As height is more important to aircraft safety than any other aspect of the 'plane's performance, the most critical instrument is the altimeter (which measures it). The evolution of the altimeter provides a compelling case study of information design and was one of the first cases where technological progress actually made ergonomic intervention necessary.

The problem with the altimeter was that when aircraft only flew up to 10,000 feet it was possible to have an altimeter with one pointer which could encompass all the necessary information in one 360° sweep. As aircraft ceilings increased to 40,000 feet it became necessary to have more information and the three pointer altimeter evolved. With three pointers of different lengths, the pilot had to do an additional sum to work out his height. By the end of the Second World War there was evidence that a substantial number of pilot error accidents were caused by misreading small angular displacements on the three pointer altimeters.

A single sweep altimeter from the First World War.

A three pointer altimeter from the Second World War.

A contemporary altimeter.

Some civil aircraft accidents caused by these mis-readings gave rise to an official altimeter committee which was set up by the Minister of Transport in 1959. Going back to a proposal made as long before as 1945, the committee suggested that all future altimeters should have not only a dial with pointers (which can show *rate of change* and *proportion*), but should also have an open window numerical display giving the height in numbers.

The problem with any numerical display is that to be understood, it requires an intellectual process to take place. This incurs a time penalty which creates more scope for error.

A new generation of cockpit instruments with numerical displays on CRTs or LCDs* re-opens this problem of legibility.

The most familiar demonstration of the ergonomic problem of legibility comes with the displays on timepieces. The traditional analogue clock face clearly shows proportion and change, but the digital display is more precise. The analogue face can be interpreted instantaneously, the digital display requires a time-consuming intellectual exercise to take place.

The best way of presenting data on a visual display has not yet been determined: "European standards for visual displays require character sizes of at least 2.6mm. The United States has no such standards."

*CRT=Cathode-ray tube
LCD=Liquid crystal display

Just as increased ceilings of aircraft brought about changes in altimeter design, so other technological advances demand new solutions from ergonomists. A Second World War fighter had about twenty instruments, a first generation jet had about fifty, but the current F-16 is so demanding and complicated that it is unflyable without a computer and, besides his analogue dials and digital displays, the pilot also has five CRTs giving him hugely variable information on a need-to-know basis. The Boeing 707 had about 180 warnings for the pilot; the current 747 has more than 450!

Rapidly improving technology has influenced the pilot's environment as much as any other aspect of aircraft. Compare the complexity of the instruments and controls of the First World War Sopwith Camel, 1; with its military and civil successors: 2 · Hawker Hunter (ca. 1955). 3 · General Dynamics F-16 (ca. 1980). 4 · Boeing 707 (ca. 1960). 5 · Boeing 747 (ca. 1980).

To solve the organisational problem there is an international convention that aircraft instrument panels be grouped into *short term* and *long term* areas. The short term instruments are the primary flight information which, because they need constant scrutiny, are in the pilot's line-of-sight. The longer term instruments are placed to the side. The head-up-display takes advantage of this hierarchy. Because too much information can overload the pilot, variation in the intensity of illumination can be used to influence the pilot's interpretation of the signals he receives.

The pilot's environment has to be *organised*, so as to minimise the risk of error caused by inattention or by over-action. The Institute of Aviation Medicine has determined a hierarchy of warnings:

1. Killer   mandatory immediate response
2. Warning immediate attention & action
3. Caution   immediate attention
4. Status   be aware

The Head-Up-Display (HUD) is an aerospace innovation which will have a certain influence on passenger cars in the future. Particularly useful in low level flight, the HUD allows the pilot to get data without having to 'take his eyes off the road'.

New display technology will make all conventional dials and displays obsolete.

In the United States, Boeing and McDonnell-Douglas are working on computer generated colour displays which give the pilot a graphic illustration of the landscape and his flightpath. Large amounts of data can be condensed into a changing image.

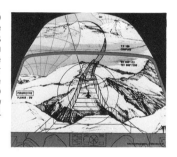

According to McDonnell-Douglas, in the cockpit of the future images will replace numbers and a graphic screen will *show* the pilot where to fly. The pilot will only need to touch the 'Big Picture' screen and to issue voice commands to fly the aircraft.

These CRT graphics will in future combine with Direct Voice Input and Artificial Intelligence to make pilots into the managers of systems rather then merely the operatives of a flying machine.

It is inevitable that the CRT (or the low power flat panel displays that will succeed it) will dominate aircraft cockpits. This same technology will be passed down from aerospace, which is at the peak of the technological pyramid, to all workplaces where displays of information are necessary: from power stations to cars.

23

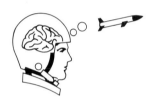

The pilot of the future will wear a 'Bio-cybernetic' helmet which measures brain activity and translates electrical impulses into commands. Within ten years McDonnell-Douglas say the Big Picture display will be able to register simple thoughts and act on them.

The lesson is an extraordinarily simple one: a picture is worth a thousand numbers. A fundamental part of the process of studying natural design is this (re)discovery of simple truths and learning to keep them in sight during the design process.

THE PORTABLE RADIO

Semantics has not only a technical meaning, in the sense of how data might be understood, but also an emotional one in the sense of what do people feel about products.

One simple truth is that people can be moved by appearances.

The portable radio is one of the most familiar consumer products and now that the technology employed in it is stable and of a universally high standard, consumers can discriminate on emotional grounds. Manufacturers are aware of this and like to test reactions to new products in 'clinics'. In a product clinic, the visual character of a new car or a new appliance is subjected to scrutiny alongside its peers.

Semantics does not only apply to the understanding of information, but also to 'softer', more humane areas such as product identity. The *meaning of* a product such as a portable radio is often crucial to its sales success.

Volunteeers are asked to assess the semantics of a design:

Is it rugged or fragile?
  soft or hard?
  weak or strong?
  friendly or aggressive?
  hot or cold?

Different markets may prefer different character-istics, but it is certain that a new product will not suc-ceed if it ignores the natural human response. In the future, designers will become more aware of this.

3. POSTURE & COMFORT

After the complex socio-psychological questions of perception, legibility and semantics, the next big problem in egonomics is the mechanical one of comfort and posture.

Although the questions "What is most comfortable?" and "What is the most efficient posture for work?" seem straightforward, their simplicity is in fact illusory because there is no precise scientific definition of *comfort,* while cultural conditioning tends to affect our attitudes to posture.

Some 'primitive' tribes relax by standing on one foot. Even in the sophisticated arena of the European motor industry, national distinctions in the attitude to comfort are apparent: French car manufacturers make seats which are seductively and voluptuously soft and embracing; Germans prefer hard and unforgiving seating. Millions of people say they are 'comfortable' on each type.

Sitting on chairs is Western orthodoxy. The Tasmanian aborigines relax by standing on one foot.

Of all traditional types, perhaps the most well-considered are the chairs which pilots use, incorporating fully adjustable lumbar and thigh supports, with positional adjustment in every plane.

This is an area where ergonomists have to advance by making careful observations and measurements.

The efficiency of a commercial pilot's workplace is perhaps the most critical in the world. The British Aerospace 125-800 pilot's seat is state-of-the-art in ergonomic efficiency.

In 1940 Walter Dorwin Teague remarked that "The automobile manufacturers have made, in the past few years, a greater contribution to the art of comfortable seating than chair builders had made in all preceding history". The Recaro 'Idealseat-C' is the most advanced car seat currently available, with pneumatically adjustable thigh and lumbar support.

Having observed that, given the choice, few people choose to work at a desk, American designer, Niels Diffrient, designed a chair (or more accurately, a sophisticated seating system) to cater for the most natural position.

Diffrient had noticed that Thomas Jefferson, Mark Twain and Ronald Reagan all chose to work in a reclining position. Winston Churchill used to do his correspondence in bed.

Niels Diffrient (1928–) one of the most outstanding exponents of ergonomics as an aid to design.

Throughout history, many successful people have preferred to work in a reclining position.
During his visit to Cairo in 1951, Persian premier, Mohammed Mossadeq, gave formal interviews from his bed.

Reclining positions appear to be more efficient (as well as more comfortable) for all sorts of tasks. In the early Fifties the Royal Aircraft Establishment converted some 'planes so that the pilot could operate from the prone position. In the Sixties, racing car designer, Colin Chapman, decided that his drivers could exercise better control if they reclined, rather than sat close to the steering wheel.

Diffrient had built up a vast knowledge of posture while working for the New York consultancy of Henry Dreyfuss where he was responsible for the interior design of Lockheed aircraft. He decided to make a seating system that was a genuine response to the problem of sitting and working.

Just after the Second World War the Royal Aircraft Establishment converted a Gloster Meteor to test a prone position for the pilot. The experiment was not continued.

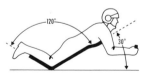

The evolution of the racing car has seen drivers adopt increasingly reclined positions. Not only does this reduce the frontal area of the car, but also offers more delicate control. 1·Juan Manuel Fangio at the wheel, (ca. 1950). 2·Jim Clark at the wheel (ca. 1967). 3·Nelson Piquet at the wheel (ca. 1985).

A part of Diffrient's approach is to disdain fashion and to search for natural principles. He says:

"Product design has been drifting more and more towards the art side and less towards the practical side ... my approach is that you earn the right to add the decorative aspects to the design after you have made it work, made it fit the job to be done, not before".

He called his new chair the 'Jefferson' because after a visit to Monticello he discovered that the old President liked to work in an arm chair, with a foot rest and papers spread out on a task table.

The Jefferson chair employs an ingenious, but simple, mechanism which lets the user adjust it almost infinitely. It is suspended from a pivot point which is close to the *natural* pivoting point of the body.

The chair is hard, with only a very superficial covering of foam because Diffrient believes that a soft chair is, paradoxically, hard work because as the limbs sink into the upholstery, muscles have to be engaged to maintain posture.

Niels Diffrient's 'Jefferson' chair will probably go into production in 1986. It is a fundamental rethink of the working chair.

Thomas Jefferson's own chair in his house at Monticello.

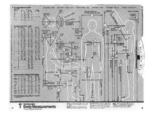

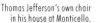

The New York design consultancy of Henry Dreyfuss, where Diffrient was once a partner, has specialised in ergonomics. Their 'Human Scale' publication was the result of years of research.

4. SPACE ORGANISATION

Comfort and posture are just two aspects of the larger question of the organisation of space. Whether at work, in a car or at home, our physical relationship with the tools and instruments we use is of fundamental importance both to comfort and to safety.

Office workers now sit at desks with visual display terminals (VDTs). The entirely paperless office is not yet with us, but it is getting nearer. Instead of ledgers, clerks now stare into electro-luminescent screens.

But it is a bastard technology, where a VDT* (derived from 50s television technology) replaces paper and a computer keypad (derived from typewriter technology) replaces pens. This mixture of technologies causes special problems for the user and the ergonomist.

The typewriter technology of the first generation computer keypads will eventually be replaced by moving 'mice', by 'touch-screens' and by Direct Voice Input.

*VDT=Visual Display Terminal

The first generation office computers used 'batch' data entry with punched cards, but the advanced VDT allows a sort of continuous electronic 'conversation'. In the jargon, this was the arrival of 'interactive' computing.

But while great benefits in efficiency were brought about by the revolution, the new work practice broke a number of natural laws. Office workers who use VDTs persistently complain of eyestrain, backache and fatigue. The *Harvard Medical School Health Letter* has reported that VDT users report sick more frequently than other office workers (although there is some evidence that similar discomforts afflict conventional office workers, who might just be less willing to vocalise their grievances).

The problems are eyestrain caused because the eye of a VDT operator is normally 25 inches from the screen, creating more effort in focussing than the 15 inches of normal clerical work; glare on the screen, from reflected light; a cramped and restrictive posture because the fixed VDT requires a fixed gaze (while traditional office techniques give workers a degree of freedom to shuffle papers); fatigue caused by competition with the tireless electronic brain.

Although in Sweden the reaction has been to limit by law use of a VDT to four hours a day, there are more sensitive and intelligent solutions than just enforcing longer rest periods.

Although IBM has had a 'human factors' laboratory since 1952, it is only very recently that manufacturers are being forced to understand the urgent necessity of designing more comfortable and more efficient work-spaces as they encounter consumer resistance to poorly designed products.

Technology will loosen-up the procedures. Already, Apple's 'mouse' and computers which have a 'touch screen' display liberate VDT users from slavish attention to the keypad. Texas Instruments computers have Direct Voice Input (DVI).

But the keypad will likely remain the 'critical interface' between man and machine. Its design has not moved on much since Eliot Noyes carefully sculpted the IBM Selectric typewriter on ergonomic principles in 1962. Recent technology has removed mechanical linkages and replaced them with electronic signals. This brings advantages of speed and means that keys cannot jam, but it also removes the sensory feedback which experienced typists need to govern their performance. One solution has been to introduce an artificial 'click' and to build in artificial 'resistance'.

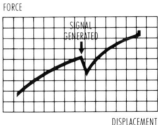

FORCE

SIGNAL GENERATED

DISPLACEMENT

Now that mechanical linkages have been removed from keypads, engineers have to build-in artificial 'feel'. Research has shown that typists anticipate a certain resistance when they press a key and become disorientated if they don't perceive it. This curve shows the ideal relationship of force and displacement.

Ergonomists have also determined that the whole keypad should be divided into 'touch' and 'no touch' areas, so that critical functions such as 'Cancel' and 'Cut' cannot be invoked without a conscious decision by the operator: they are put outside the natural work area.

Keypads are now designed so that the familiar 'QWERTY' keys are in accessible areas, but the more crucial commands such as 'CANCEL' and 'EXIT' are in positions which require a deliberate action to implement.

But how should the whole working space be organised? Some research has shown:-

1. The VDT should be at least 14 inches on the diagonal and should tilt from $-5°$ to $+20°$ swivelling $45°$ left and right. The screen should be etched matt for glare control.

2. The sitting position should not impose surplus weight onto the thighs so that blood flow is uninhibited.

3. The angle between the trunk and the thighs should be more than $90°$.

4. Arm and wrist rests should be provided.

5. There should be fidget potential designed in, so that VDT operators can alter their posture without impairing their work.

6. The backrest of the seat should follow and respect the natural curvature of the spine.

7. The seat should not be so soft that its occupant has to use muscle power to locate himself.

8. There should be a footrest.

9. It should be easy to get into and out of.

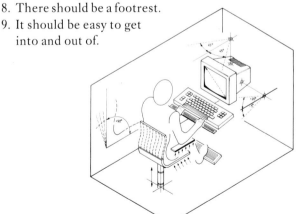

The 'ideal' VDT workstation.

The design of the workstation in a car is a, perhaps, more familiar problem. The major ergonomic questions in vehicle design are the efficient organisation of the driver's environment, and the reduction in the number of controls and an increase in their simplicity.

These problems are essentially the same as those that confront the aircraft designer because in all vehicles ergonomic research is restricted by its application to a dynamic environment. In cars the problems are specially acute because one mass-produced design has to fit an almost limitless variety of people.

The chief problem in contemporary car design is the variability of the human frame: people differ in size, reach, strength, as well as in their speed and accuracy of perception and their coordination. A sophisticated car interior, such as the Ford Scorpio, has to cope with all the variations and yet be acceptable to each individual. With the Scorpio, close attention to ergonomic detail has the happy result of creating an effect that is aesthetically pleasing.

The detailed solutions which car designers find for these problems are among the best available versions of the discourse between man and machine.

Nature, tradition and law all effect man's disposition with the machine. There are some facts of natural design whose origin no-one can determine, but whose significance is clear for ergonomists.

The Wright Flyer of 1903. The world's first heavier-than-air; 'plane had the pilot sitting to the left of the engine. This disposition reflected a certain tradition that is still not broken: cavalry mount their horses from the left and airline captains command from the left seat.

Horses can only be mounted from the left; the Olympic circuit runs from left to right; perhaps because the Wright Flyer had the engine on the right, the pilot on the left, today's airline captains command from the left seat; in an emergency, pilots almost always turn left.

The position of the man in charge of an animal or a machine is clear, but how best does he control his charge?

In cars the steering wheel evolved very quickly from the tiller and the development since then has been only slight, but car designers can already foresee the next development; an instrument binnacle· actually attached to the adjustable steering wheel, with subsidiary instruments clustered on a flexible gooseneck.

The steering wheel evolved from the tiller used on primitive automobiles, but soon achieved stasis, changing only in size and materials. Future developments include instruments being attached to the wheel in a discrete binnacle.

The Ford Eltec is a fully functional concept vehicle, designed to maximise interior space. It suggests the way in which passenger cars will develop in future with an increasing commitment to passenger comfort, convenience and safety.

In aircraft, rapid advances in solid state controls have made the old control yoke/joystick redundant and in new generation 'planes such as the A320, a sidestick control will respond to minute deflections of the muscles in the captain's wrist.

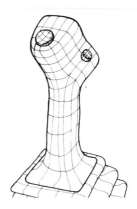

The sidestick controller from the new A320 Airbus.

These simple and effective sidestick controllers used to be unacceptable because of the bulky control runs which the old hydraulic and mechanical systems would require, but the new 'fly-by-wire' technology has brought about a fundamental change in the design of machines: *input controls can be remote from mechanisms*. This promises another revolution in human behaviour as profound as the one that occurred at the end of the nineteenth century when the generation of electricity meant that power could be distributed.

Now it is the same with *control*.

In a curious way this new technology brings the problems of *natural* design into sharper focus. When a whole airliner's safety depends on subtle deflections of someone's wrist, the control that wrist operates had better be comfortable and efficient.

The design of a cooker hob is less daunting than the creation of a safe and efficient pilot's environment, but it is no less an exercise in ergonomics.

Searching for an arrangement of controls to burners which offers fewest opportunities for error, ergonomists have discovered one layout which is superior.

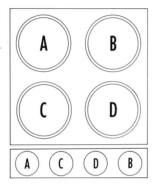

The most efficient relationship between control knobs and burners is a homely subject much studied by ergonomists.

But they do not know whether there are any specific
natural laws which these layouts follow.

Ergonomics is deceptively simple. Deceptive because it is the simple lessons that are most difficult to learn. Observation and practice suggest to designers certain rules about how machines should be designed with people in mind.

Seats should support you; controls should be in reach; data should be stored in a selective hierarchy; knobs should suit the hand's precision grip between thumb and forefinger.

Like the steering wheel, some familiar products have reached stasis. The electric shaver evolves only slightly year by year, but car radios, under the influence of burgeoning technology are continuously getting more responsive to the user's needs.

Because they have developed over a long period and because the technology itself has become relatively stable, designers of electric razors have been able to develop shapes which are very efficient in terms of ergonomics. The earliest razors made few concessions to the natural design of the hand (1951, 1963, 1971), but accumulating experience with this most elemental of functions over a production run that has reached hundreds of millions has given Philips, the market leader, data to produce designs that have genuine ergonomic properties (1985): pleasant to hold, easy to use and good to look at.

The car radio should be a critical essay in ergonomic design, but is too often ignored. Because the driver's attention is limited, a car radio must be specially clear and easy to use. Its design must be foolproof and should be coherent and legible without being distractingly assertive. The latest Philips radio for the next generation Austin-Rover car has taken full account of current thinking in ergonomics.

The questions of ergonomics are simple ones:

Is it safe?
Is it comfortable?
Is it efficient?
Is it robust, reliable, convenient and attractive?

But the answers are altogether more variable and complex.

Take the British Aerospace Hawk and a future project for a camera by Luigi Colani. The Hawk followed to the letter the provision of AFP970, the 'Bible' of aircraft ergonomics, and it became the most successful British 'plane of modern times.

Colani's camera has yet to go into production, but, like the Hawk, it is a response to the very specific requirements of making a machine which complements the human body and does not oppose it.

The study of ergonomics does not presuppose identical solutions to similar problems: the British Aerospace Hawk is the most successful British military aircraft of modern times. Its design followed almost to the letter the provisions of AFP970, an official 'ergonomic' specification. On the other hand, Luigi Colani's "bio-designs" for Canon derive their inspiration from fish, animals and trees.

Natural design is the business of rediscovering these simple principles. Like all that is best in any field of activity, natural design is proof of man's ability to transcend his limitations.

References

Malcolm Ballantine "Conversing with computers – the dream and the controversy" *Ergonomics* vol. 23, No. 9., pp. 935-945, 1980.

Cyril Burt *A Psychological Study of Typography* Cambridge University Press, 1959.

Henry Dreyfuss *Designing for People* Simon & Schuster, New York, 1955

R. L. Gregory & E. H. Gombrich (editors) *Illusion in Nature and Art* Duckworth, 1973.

Human Factors of Workstations with Visual Displays IBM Corporation, San Jose, 1984 (3rd edition).

B. H. Kantowitz & R. D. Sorkin *Human Factors* Wiley, New York, 1983.

A. C. Mandal "The Seated Man: theories and realities" *Proceedings of the Human Factors Society,* pp. 520-524, 1982.

E. J. McCormick & M. S. Sanders *Human Factors in Engineering and Design* McGraw-Hill, London, 1982.

Walter McQuade "Easing Tensions Between Man and Machine" *Fortune* 19th March, 1984, pp. 48-56.

R. D. Ray & W. D. Ray "An analysis of domestic cooker control design" *Ergonomics* vol. 22, No. 11, pp. 1243-1248, 1979.

J. M. Rolfe "Human factors and the display of height information" *Applied Ergonomics* pp. 16-24, December 1969.

Brian Shackel (editor) *Applied Ergonomic Handbook* IPC Science & Technology, Guildford, 1976.

Harold P. Van Cott & Robert G. Kinkade *Human Engineering Guide to Equipment Design* American Institute for Research, Washington DC, 1972.

Earl L. Wiener & Renwick E. Curry "Flight-deck automation: promises and problems" *Ergonomics* vol. 23, No. 10, pp. 995-1011, 1980.

Photographs

Illustrations

Picture research

Marina Cantacuzino

Praise for *Your MBA with Distinction*

This book is a 'must buy' for all aspiring MBA students as they seek to maximise the return on the many differing investments they have to make in order to participate successfully in leading MBA programmes. It provides a practical, realistic understanding of the processes involved in undertaking and achieving an MBA degree. The work is an outcome of the two author's practical experience of both sides of the academic fence, being both MBA students and tutors at one of the UK's leading management schools. This position enables them to empathetically lead the student through the myriad of situations they will encounter during their MBA experience in order that they might gain insight and preparation that should stand them in good stead. Many important aspects are discussed such as Learning in teams, Cultural issues and Life balance which the mature student might be facing for the first time as well as some very useful hints on how to be successful in exams, assignments, presentations and projects. The overall aim of the book is to provide a recipe to assist the student in understanding and achieving the tasks necessary to successfully complete MBA studies. As such, it provides a comprehensive and accessible reference work for students and it is also an eminently sensible addition to the pre-MBA reading for all would-be students. I commend it to you.

Mike Jones, Director General
The Association of MBAs, London

This book is a highly practical guide for busy MBA students, which can help them make effective use of scarce time without sacrificing the quality of their work. The text is highly accessible but never patronising. The authors have the advantage of having studied successfully for their own MBAs before becoming MBA tutors and examiners. The book truly benefits from their ability to see what the world looks like on both sides of the MBA fence – it should be on every new MBA student's reading list.

Professor Rosemary Deem, BA Hons, Social Sciences, MPhil, PhD
Director of Teaching and Learning, The Graduate School of Education,
University of Bristol

This is a highly useful guide to the practice of doing an MBA – and from a very reliable source, since both authors achieved distinction in an MBA themselves while working full-time at demanding jobs. A serious MBA stretches everybody who does one – not only leading to deep knowledge about organisations and how they function, but also to a development of self-understanding and management capabilities. This books gives an excellent practical guide to how to go through this process and achieve the knowledge and development which a good MBA provides. If you want to make the serious investment of both time and money that an MBA entails worthwhile, this is the book for you.

Professor Stephen Watson
Principal, Henley Management College
Chair, Association of Business Schools

YOUR
MBA WITH
DISTINCTION

YOUR
MBA WITH
DISTINCTION

**Developing a systematic
approach to succeeding in
your business degree**

CAROLINE GATRELL AND
SHARON TURNBULL

FT Prentice Hall
FINANCIAL TIMES

London · New York · San Francisco · Toronto · Sydney
Tokyo · Singapore · Hong Kong · Cape Town · Madrid
Paris · Milan · Munich · Amsterdam

Pearson Education Limited
Head Office:
Edinburgh Gate
Harlow CM20 2JE
Tel: +44 (0)1279 623623
Fax: +44 (0)1279 431059

London Office:
128 Long Acre
London WC2E 9AN
Tel: +44 (0)20 7447 2000
Fax: +44 (0)20 7447 2170
Website: www.business-minds.com

First published in Great Britain in 2003

ISBN 0 273 65667 8

British Library Cataloguing in Publication Data
A CIP catalogue record for this book can be obtained from the British Library.

10 9 8 7 6 5 4 3 2 1

Typeset by Pantek Arts Ltd, Maidstone, Kent
Printed and bound in Great Britain by Biddles Ltd, Guildford and Kings Lynn

The Publishers' policy is to use paper manufactured from sustainable forests.

Caroline Gatrell

For Tony, Anna and Emma
and my parents, Pam and Max

Sharon Turnbull

For Edwin and
my parents, Audrey and Reg

Contents

Part II: A systematic approach for managing your coursework

Acknowledgements

Sharon and Caroline would like to acknowledge the contribution made by our colleagues at Lancaster University Management School, not only in connection with this book, but also in relation to our personal learning and development since we joined Lancaster University.

In particular, we would like to thank John Mackness for his guidance, not only as Director of Lancaster's Executive MBA at the time when we were students, but also as Head of Management Development Division, where we both began our University careers. Both of us would like to thank our personal tutors from our years as MBA students – in Caroline's case, David Brown and, in Sharon's case, Ged Watts. David and Ged saw us both through our MBAs and have backed and encouraged us ever since.

In addition, we would like to say 'thank you' to Val Goulding for all her assistance and kindness over the years, and to Julia Davies for her wise counsel and support.

We would like to acknowledge the contribution made by the original sponsors of our MBA studies: Mike Jones (Sharon) and Richard Crail (Caroline) – without them, we might not be here now! Mike Jones is now Director General of the Association of MBAs, and we would like to thank both him and Peter Calladine at the Association for their enthusiasm for this book since its inception.

Thank you also to Amelia Lakin from Pearson Education for the insightful suggestions she made to our earlier drafts, and to Carole Elliott and Nick Rowland, each of whom spent valuable time reading through the final draft of the book and made helpful comments.

Finally, a special thank you goes to Tony (from Caroline) and to Edwin (from Sharon), for their support and patience as we produced this book!

Copyright material

We are grateful to the following authors and publishers for giving us permission to reproduce copyright material:

Chapter 2, Figure 2.1: Gant chart reproduced with kind permission of Mark Lee.

Chapter 8, Case Study Exercise: 'Face Value: Where's the beef?' *The Economist* 3.11.2001.

Chapter 10, Figures 10.1 and 10.2, Brassington, F. and Pettitt, S. (2000), *Principles of Marketing*, Financial Times Prentice Hall, with thanks to Pearson Education.

Foreword

For most people, undertaking MBA study is a step into the unknown, and *the unknown* is always an intimidating phenomenon. It is wise, therefore, to be well-informed about MBA study prior to arrival at business school so as to avoid unpleasant surprises. MBA students have been known to compare the first few weeks at business school with being thrown into the deep end of the swimming pool by an instructor who hasn't yet bothered to tell them how to swim. The medium in which the swimmer and the MBA student find themselves may differ, but both scenarios can result in similar consternation, anguish, ineffectual thrashing around, general panic and a feeling of helplessness. This is par for the course, some might say, and after all, few of those non-swimmers thrown into the deep end actually drown. Many of them will become excellent swimmers and many MBA students, who struggle in the first term, will go on to be hugely successful in the boardroom. So what's the big deal?

The material in this book is *essential reading* for any prospective MBA. You do not have to be hell-bent on getting an MBA with a distinction to profit from this book. Success can be measured in many ways: promotion; personal satisfaction; a new career in a new industry; remuneration or, in academic terms, gaining an MBA with distinction. To be awarded an MBA with distinction is satisfying as it is academic recognition of time well spent and to good purpose. In the case of a sponsored student it also vindicates the investment made in you by your employer. In all cases, it would be hoped that the serious student would want to do well in his/her studies. This book aims to give practical advice on how to do exactly that. To gain an MBA with distinction is a laudable goal, but irrespective as to ultimate goals, anyone intending to study for an MBA would greatly benefit from the practical advice offered here by Caroline Gatrell and Sharon Turnbull.

As Educational Services Manager at the Association of MBAs, I have, over the years, advised thousands of prospective students and have listened to the experiences of many current students and graduates. I do not have a problem in pointing enquirers in the direction of useful publications on MBA study, especially to those which provide general advice on MBA study and how to select a good school. I am involved in producing one myself, *The Official MBA Handbook*. For those who have been away from formal learning for some time, which is the case with many prospective MBA students, there are also books on study skills. However, there has not hitherto been a good book on how to get the most out of MBA study in terms of obtaining good grades and a distinction. This publication is, therefore, a welcome addition to the MBA bookshelf.

It is often quoted that *'genius is 99% perspiration and 1% inspiration'*. The good news is that to gain a distinction you do not have to be a genius, although it does help. You will certainly need the perspiration and some inspiration, but this may not be enough. This book emphasises a systemic approach based on a thorough understanding of *the system*. I find it interesting that in the week that I write this foreword I read a report quoting a recent study involving medical students which found a direct correlation between the inability to meet a deadline for submission of a short questionnaire at commencement of study, and examination failure. How much easier it would be if the failures had taken the trouble to better understand what was expected of them and realise the importance of being well organised. This is not a difficult goal to achieve, it requires no great intellect, but many prospective students are so focussed on the excitement in 'going to business school', and the prospect of having to tackle new subject areas, that they completely overlook the practicalities of postgraduate study. This may be because, as a graduate they believe that they have 'been there, done that'. Whatever the reason, it is a grave oversight.

Postgraduate study is quite different from the undergraduate experience. It would be arrogant and ill advised to assume that you know what it is all about. To be forewarned is to be forearmed. Students who prior to commencement of study invest time and effort in finding out what is involved in their MBA programme will avoid much stress and get more out of the experience than the unenlightened. They are more likely to succeed and to be appreciated by fellow students and academics.

I recommend this book to ambitious, prospective MBAs and wish them every success in their studies.

Peter Calladine
Educational Services Manager
Association of MBAs
July 2002

Introduction to the book

'During your MBA you make lifelong friends from around the world and across an array of sectors... these friends will become the senior management, investment bankers and management consultants with whom you interact for the rest of your career – an invaluable business network.'

Paul, Cambridge University

The aim of this book is to help you succeed on your MBA.

Whether you are a part-time, full-time or distance-learning student, *Your MBA with Distinction* will de-mystify the secret of getting an MBA and help you improve your performance. The authors believe that gaining an MBA is a *process* which, if approached systematically, can be managed effectively. Caroline Gatrell and Sharon Turnbull share a unique understanding of what is required to negotiate the difficult pathways of an MBA, and these 'secrets of success' will be explained to you. By adopting the methods of study and approach described in this book, you should be able to rapidly enhance your level of achievement on your MBA programme or in any business degree. The advice in this book will also assist you in saving precious time along the way.

We have used the phrase 'with distinction' in the title of the book, since this implies both high achievement and excellence. Some MBA courses, particularly many in the UK, award a 'distinction' grade for those achieving the top marks in their business degree. Whilst we recognize that grading systems vary considerably from business school to business school, the purpose of this book is to support you in maximizing your success on the programme. We do hope that many of you will earn a 'distinction' or the equivalent high grade in your MBA as a result of reading this book. We have inserted 'tips for excellence' throughout the text, and if you are aiming for a top grade we recommend that you pay particular attention to these.

Purpose of this book

The purpose of this book is to:

- Share with you the unique knowledge of the authors, both of whom gained MBAs with distinction, and both of whom are now MBA examiners.
- Provide an essential, step-by-step guide to the process of how an MBA works.
- Help you improve your grades.
- Enable you to manage your own learning effectively.
- Help you to 'take control' of your MBA, rather than feeling submerged by it.
- Help you to improve your time management.

1.1 Doing well in your MBA

The key to success in your MBA can appear to be shrouded in mystery. At first, it may seem as though 'excellence' is a concept available only to individuals who are either very lucky, or brilliant. The authors of this book have found that even the most gifted students can have difficulty in understanding what is required to optimize their performance in an MBA. This means that they often go through a long and painful process trying to understand what is required before they get themselves 'on track', meaning that valuable time and marks may be lost. This text will enable you to make the most of your abilities, as well as the time you have available, while developing your individual learning and writing style.

The book will also include some handy *Tips for excellence!* for those who seek to excel in some part or all of their programme. The tips for excellence – which focus on the little 'extras' that can help you gain a distinction or an excellent grade, are concentrated in those chapters which are about preparing work which will be assessed by your university. Each section in the book is clearly set out and cross-referenced so that you can use it as a trusted guide in which it is easy to locate advice quickly when you need it. Chapters do not have to be read 'in order' – you can dip in and out of the text to suit your own needs. Real-life examples are given to illustrate points (though names and specific details have been changed for reasons of confidentiality). The use of jargon is avoided in order to make the concepts accessible, but a short glossary of relevant management and academic terms is provided at the back of the book.

The book focuses on the hurdles that all students (whether full-/part-time or distance-learning) will face:

- writing assignments
- taking exams
- working in groups
- doing case studies
- writing a dissertation
- choosing electives
- managing the difficult times
- applying theory to 'real-world' problems
- making successful presentations.

The authors have also focused on the needs of particular groups, with chapters on

- managing distance- or e-learning
- making the most of classroom learning.

1.2 The authors as both MBA students
and MBA teachers

Your MBA with Distinction is derived from the authors' own experiences of teaching MBA students and of previously undertaking MBAs successfully themselves. The writers, Caroline Gatrell and Sharon Turnbull, were both practising managers when they gained their MBAs (both with distinction) from Lancaster University's five star[1] Management School. This means that they have experienced the MBA *from the student point of view*, and that they understand

---

[1] Top rating in UK research assessment exercise.

completely what it means to succeed on an MBA while combining study with full-time work. As well as their student perspective, however, Caroline and Sharon now have an excellent academic insight into what is required for an MBA, since both have now joined the faculty of Lancaster University Management School, where they are working at a senior level as teaching fellows. This means that the authors have been extensively involved with world-class MBA programmes, both full-time and part-time. They therefore have a unique perspective on how to succeed on an MBA, each having sat MBA examinations as candidates, but both now in the position of setting and marking MBA exams as examiners! The same applies to assignments, dissertations, and presentations – the authors know what it feels like to be both student and teacher. Having 'been there' themselves, Caroline and Sharon know only too well which aspects of the MBA students find most difficult, yet as academics, they are in a position to provide good, clear guidance as to what is required, in the most effective and least painful way possible!

1.3 A guide to what is in the book

Part I: A systematic approach for managing the learning experience

Your MBA with Distinction is presented in two parts, to provide greater cohesion for the reader. *Part I* gives general advice on how to manage the 'learning experience'. For example, many of you will have to learn to work successfully with fellow students from different cultural backgrounds and, if you are an overseas student, learn how to succeed in an institution based in a country which is not your own. Advice on this and other general issues is given in Part I.

Chapters 2–7

Chapter 2 is devoted to the development of *improved study skills* (including literature searches and selective reading) and the concept of self-management. Advice on effective listening and note-taking, time management and exam revision is provided. This takes account of the needs of all students: full-time, part-time, and 'e-' and distance-learning.

Chapter 3, on *sharing knowledge and being supervised*, explains what tutors/faculty advisers expect from students and how students can make the best use of whatever supervision is on offer. The difficulties and benefits of working in learning sets/tutorial groups (including online groups) are discussed and students are given suggestions as to how they might enhance their own learning by working in collaboration with (as opposed to in competition against) other students. Working across cultures, effective communication and the management of group projects are discussed. Real-life (but anonymised) examples will help illustrate these points.

Over any period spanning between one and three years, most students will experience a 'low' period. This may be due to factors directly related to the MBA programme or the intrusion of life events, such as divorce or bereavement. Chapter 4 gives *suggestions about how 'low' periods may be coped with* in the context of part-time, full-time and distance-learning study.

Whether students are in the classroom for a year as full-time MBAs, or only a week during the summer as distance-learning students, it is vital that they make the most of *the classroom environment*. Chapter 5 deals with how to get the best out of what is on offer – how to contribute to group discussions and how to manage when you don't understand what is being taught in the classroom.

In Chapter 6, attention is given to those students who are undertaking their MBA *'at a distance' and are reliant on the e-environment* for a large part of their course contact. As well as considering the advantages and disadvantages of the virtual learning environment, this chapter tackles the general difficulties faced by students who are learning 'at a distance', and gives advice on how to manage these.

Chapter 7 discusses what criteria students might use for *choosing electives*, both in the short term (doing well in the MBA) and in the longer term (enhancing career prospects).

Part II: A systematic approach for managing your coursework

Part II provides specific guidelines on managing coursework. Despite some differences in style and emphasis (and whether full-time, part-time or distance-learning), most MBAs are taught in English and share the same basic curriculum, originally derived from the traditional American model. This book draws together the 'common threads' that run through most MBAs and offers guidance on issues that many MBA students will face. For example, there is a significant difference between what is regarded as a good, written report in the workplace and what makes a good assignment/dissertation at university. This can cause high levels of anxiety on the part of MBA students, who may lose valuable marks because of it. Part II, Chapters 8–13, covers the key areas of knowledge required to succeed in MBA coursework.

Chapters 8–13

In Chapter 8, guidelines are given to assist students in *case study work*. These take account of the fact there may be various approaches to case studies, ranging from traditional classroom teaching, to more informal small groups who may even be working online. A practical example is included.

At some point in their MBA programme, students will be expected to take their learning out of the classroom and into the workplace in the context of faculty-supervised field studies. This may be particularly pertinent to distance/online and part-time MBA students whose courses are probably designed to incorporate day-to-day work practice with study. However, it is also relevant to students on more traditional full-time MBA programmes, who will be encouraged to conduct field studies in small teams, working closely with the sponsoring organization and a tutor/faculty adviser. Lancaster University Management School specializes in helping students *apply business theory to real world problems* and, in Chapter 9, readers are advised how best to relate their learning to study in the field.

Chapter 10 focuses on how to manage the process of *producing a good assignment*, whether this is based purely on literary concepts or on 'live' field studies. Guidance is given on how to apply theory to practice. This section will include a clear explanation of what is expected in an assignment, and why. An insight into the mind of the marker is included. Sample assignment plans are provided to give you an idea about what you should include in your assignment, as well as how you should structure it.

Presentations are often a compulsory component of assignments or of the MBA dissertation. Students often find presentations difficult – not only because they are nervous about speaking in public, but because the requirements of an academic panel can be different to those of colleagues in the workplace. Chapter 11 offers advice to help students overcome these fears and give them a better understanding of academic requirements.

Chapter 12 aims to help students re-position the concept of the *exam* in their thinking, from an event over which they have little control, and in which luck plays a significant role, to a process which can be managed successfully. Techniques which will assist you to pass exams, as well as advice on calming exam nerves, are provided. Real-life examples are used to illustrate the points made and exam practice exercises are suggested.

Chapter 13 focuses on the *dissertation* – a long haul at the end of the MBA which can make a significant difference to your overall marks, as it is often weighted heavily in comparison with other parts of the programme. Guidance on managing the dissertation process (whether within the workplace, or as a 'research' process), will be given. A significant part of this chapter will be devoted to research design – choosing a relevant, feasible topic, and the choice of an appropriate methodology. Consideration is also given to the politics and practicalities of conducting organizational research, including ethics and career enhancement.

A brief *glossary* of popular MBA and management education 'speak' and *references and suggestions for further reading* are given at the end of the book.

1.4 A note about what is *not* in the book!

Before you begin reading it, it is probably worth noting what this book does *not* intend to do.

Selecting an MBA programme

The book is not about selecting an appropriate MBA programme for your needs. It begins with the assumption that you have made your decision and are either about to embark on or are already engaged in your MBA programme. If you are still considering which MBA programme(s) to apply to, please consult the further reading section at the end of this book for additional guidance.

Revision aids

The book is *not* intended as a revision aid, or 'crammer' because the authors do not believe that successful MBAs are obtained by 'cramming' facts just before the exam element of the programme. MBAs should be approached in a rounded, systematic manner, in which time and learning are well managed. However, for those of you who do feel you need to refer to 'crammers' just before your exams, there is a good selection on the market and some suggestions are made at the end of this book.

Investing yourself in your MBA – the book cannot do this for you

If you follow the guidelines in this book, you can make a difference to how well you do in your MBA. But a book cannot obtain an MBA on your behalf. The amount of time and effort you give to your course will also have a direct impact on how well you do. As a general rule, what you get out of an MBA, and the level of your grades, will only increase as a direct proportion of what you put in. If, for one reason or another, you don't put a lot into your MBA, then you won't get a lot out of it!

The book should not supersede your own faculty guidelines

There are now hundreds, if not thousands, of MBA programmes running throughout the world. Most of these will follow the same core curriculum, but *each will also have its own institutional regulations* about what is expected from students in terms of coursework.

Guidelines relating to submission of work, word limits, deadlines and so on will vary. It is very important that, as well as reading this book, you take note of the requirements of your own faculty and listen to the counsel of your own faculty advisers. Whatever your own institutional regulations advise, you should follow them – and they should take priority over the general advice given in this book.

Read on, and enjoy your MBA!

Part I

A SYSTEMATIC APPROACH FOR MANAGING THE LEARNING EXPERIENCE

Study skills

'I found that the key to successful completion of the many essays, cases and assignments, was to find a new daily and weekly routine and stick to it. I soon got into a new pattern, and even managed to keep a reasonable social life going as well!'

Rick, Erasmus University, Netherlands

Introduction

For many, embarking on an MBA means returning to studies after a long gap. For others, the MBA may be a natural progression after having recently completed another qualification. Whether your experience is recent or distant, however, we believe that it will be worthwhile to review your approach to studying and plan for this particular programme before you start.

The intensive nature of the MBA, and the fact that many of you will be studying part-time whilst continuing to hold high-powered professional jobs, means that thinking about how you will approach the programme will be time well invested. If you have chosen to pursue your MBA on a full-time basis, you will find that the pressures are no less great. Full-time programmes are conventionally shorter in duration, and business schools feel that they must expose you to an extensive repertoire of knowledge in the time that you are with them. So, whichever scheme you have selected, you will find that once you begin an MBA there is little time to pause for breath.

Over the years of working with MBA students, we have come to the conclusion that good planning, effective time management, and a methodical approach to studying can all make a huge difference to the grade achieved. No less important, we have also found that the quality of life students experience whilst studying for this exciting but demanding qualification can be enormously affected by how well they prepare for the experience.

Purpose of this chapter

The purpose of this chapter is to:

- Offer you a guide to preparing for your studies.
- Help you to rebalance your life in preparation for your studies.
- Develop effective times and project-management skills.
- Help you to conduct productive literature searches.
- Introduce techniques for effective reading, and ways to record your findings and take notes.

If you have recently been studying, we suggest that you consider our suggestions in the light of what has worked for you, but remember that the MBA is a different experience from many other forms of study. Do be willing to try some new ideas as well as retaining some of the tried and tested methods that have worked for you in the past. Ask yourself what difficulties you experienced during your last course. How well did you succeed in fitting study into your already busy life? We hope that some of the suggestions in this chapter, all of which have been developed over years of discussions with our students, will help you avoid some of the common traps many have fallen into in the past.

If you are returning to studying after a long gap, you will find that the world of study has changed dramatically! Even if your break has been short, it is inevitable that technology will have made huge strides forward. There is now an increasing number of information resources available to you in electronic form through libraries and the internet – data that once would have taken hours, even days to source. You may be feeling daunted by the prospect of starting again after such a long gap, or by the fact that your undergraduate degree was in a subject unrelated to a business degree. Don't worry. This is a common feeling, and you will not be alone.

Studying for an MBA is likely to feel very different from any previous courses you have done. Unlike many traditional taught courses, the MBA demands a high level of self-management and active participation. Your tutors will be pointing you in interesting directions, leading you into new academic terrains, but how you respond to their lead, and how much you gain from the programme, will depend on how successfully you adopt a mindset of curiosity and tenacity. The most successful students are *hungry* for knowledge and ideas. Learning for them becomes a *quest*, a journey of *exploration*.

In order to gain maximum benefit from this type of programme, we urge you to master a small number of basic study skills that we outline below. These are designed to enable you to enter the intriguing world of management and organizational knowledge. The aim of this chapter is to help you to do this as quickly and efficiently as possible.

2.1 Getting prepared – thinking about the balance of your life

One of the first questions we always ask potential recruits onto our MBA programmes is whether they have considered the impact that this commitment may have on family, friends and colleagues. This sounds obvious, but overlooking this step has led many MBA students to feel pressured because they are torn between their various commitments. This is why we consider that setting aside some time to plan is so important.

Ten steps to rebalancing your life in preparation for the MBA

Life will be much easier during your programme if you have prepared your friends and family in advance for what is likely to be expected of you and how this will affect your lifestyle. We firmly believe that finding a balance in your life whilst doing an MBA is not impossible. However, we also know that during the programme, life *will* change and old routines are likely to be threatened unless you can find effective ways to reschedule those parts of your life that are important to you. All of the following should ideally be done before you start your programme.

Step 1

Think about the balance of your life as it is at the moment in a typical week. Calculate how much time is devoted to work, family and leisure activities. Make a note of each part of a typical week so that you can return to this later to decide how you will make the time you are going to need for your studies. You might be surprised by the calculations you have made.

Step 2

Make a note of the time involved in the *formal* content of the programme. How many hours of lectures, seminars, tutorials, and online discussions will you be committed to? The prospectus should have made this clear. If not, check this with the course tutors.

Step 3

Ask your tutors about their expectations of the number of hours of *private study* you will be required to undertake per week. All MBA programmes require a considerable amount of private study, and it is this hidden part of the programme that can be some students' downfall. You will probably find that your tutors are reluctant to give you any absolute figures as this can vary from student to student. However, as a very rough guideline we estimate that as a part-time student you will need to find at least 12 hours for private study per week and, as a full-time student, between 15 and 20 hours. You will need to manage your time as you might manage a financial loan, budgeting to spend the hours needed each week and remembering that if you do not pay in full in one week, you will need to pay more the following week, sometimes with interest if you fall behind! If you are studying your MBA in English where this is not your mother tongue, you may need to devote a little extra time on top of this, as you will not necessarily be able to read and digest the literature as quickly as native English speakers. If you are returning to studying after a long gap, you may also find that you need time to settle into a routine of studying. Don't worry – this need not last very long – you will soon be as efficient and confident as the others.

Step 4

Try to seek out some previous students from the programme. They will be able to give you a good estimate of how many hours you will need to survive, and how many more you will have to put in, in order to do well. Our advice is to factor in some 'contingency time' in case things go wrong. We have seen too many students cut the time allocated for their studies to a minimum. They can then fall behind as a result of an unforeseen short-term crisis.

Step 5

You will now have a more informed idea about how many hours you will need to set aside for your studies. Consider what you will need to give up in order to make this possible. Be realistic, and don't be too hard on yourself. It will be important for your own well-being that you do retain some of the other pleasures in life whilst you are studying.

Step 6

Plan your new routine. Think about whether you work better in the mornings or in the evenings, whether you prefer to build in a period of private study on a daily basis, or whether, if you work full-time, you would prefer to do most of your studying at weekends. Decide what times you will preserve for family, friends and hobbies. Remember to plan for holidays and breaks.

Step 7

Discuss your proposed new routine with all those affected. Seek their agreement about when you will study and when you will spend time with them. Also seek their support in not disturbing you during the times you have allocated for study.

Step 8

Find a quiet place to study. You will need somewhere quiet where you can spread your books and papers out. If there is no space at home, try to find an alternative space, either at your place of work or in the library. Find some storage for your books and files. (The husband of one of the authors spent the first term of her MBA learning woodwork in order to build her a bookcase! It served two important purposes: it provided somewhere to keep the growing pile of books and folders, but more importantly it provided a new hobby for him whilst he was temporarily taking second place to his wife's studies!)

Step 9

If you are studying for an MBA on a part-time basis, it will also be important that you engage the support of your superiors and colleagues at work, as there will be times when you are absent from work and you may need to ask them to cover for you. If your course emphasizes action-learning and the application of theory to practice, you should find that there are opportunities during the programme for you to do assignments and projects in your workplace. If you have the support of your colleagues in advance, they are likely to be much more receptive to your requests for information or time later on. If possible, try to find a mentor at work, perhaps a more senior manager who is willing to discuss your learning and its application in practice with you as you progress through the programme. If you are pursuing a full-time MBA, you may still find it useful to seek the support of a senior practitioner who is willing to act as a mentor to you. She or he may help you to understand the practical application of what you are learning and may also be able to offer project opportunities to you during the course.

Step 10

Having put all of your plans in place, do your best to stick to them, but don't panic if unforeseen events get in the way. You will be better equipped to cope with the unexpected if you have established a study routine and built self-discipline into it. If a personal or work problem crops up, it may mean that your study routine becomes disrupted for a short while and you therefore have to build extra hours into future weeks to make up for the time you have lost. Most students have to do this at some point during the course, but still successfully complete on time, so there is little need for concern. However, if the problem seems to be serious, and is likely to affect you for more than one or two weeks, alert your tutors immediately. If they know that you are experiencing difficulties they will be able to help you (see also Chapters 3 and 4).

2.2 Time and project management

There are many books written specifically on the subject of time management. Space does not allow us here to go into any depth on the various systems available for this purpose, but we do recommend that if you have not already adopted a successful way of managing your time, you should do some research as soon as you can and adopt a system that meets your needs.

We do, however, want to devote a little space to *project management*, as this is a core skill which often differentiates excellent students from the rest. During your MBA programme, you will be asked to undertake a wide range of projects ranging from short essays to work-based assignments, group projects and a major dissertation. If you approach each of these as a 'project' that requires careful planning, you are more likely to meet your deadlines and to be satisfied with the quality of your work.

There is much written on project management systems, but for the purposes of the MBA, you will not necessarily require anything very sophisticated. The use of a simple Gantt chart will suffice for most MBA projects. Figure 2.1 is a real example of an assignment project plan. The student has counted the weeks available to him and plotted each of these along the horizontal axis. He has then determined the tasks required for successful completion of the project and listed these down the vertical axis in time order. This chart has enabled him to view, in any one week, the tasks he has to achieve and the time available to him. Of course, in practice it is never possible to be totally accurate with estimates, but you will become more accurate as you become more familiar with the work required in doing assignments. A simple Gantt chart like the one illustrated, however, can be easily updated during the lifetime of a project and is an excellent way to keep yourself on track, as well as keeping your tutor updated on your progress.

2.3 Productive literature searches and effective reading

Sourcing appropriate references, reading them and taking notes are all skills which are going to be crucial to your success on your MBA programme.

As we will explain further in Chapter 3, you are unlikely to be 'spoon-fed' with readings on an MBA programme and, even if your tutors are kind enough to provide you with some readings to get you started, these should only serve as a starting point for your research. The further you go in researching the theory related to the topic you are studying, the greater will be your understanding and your grades should start to reflect this. Over the years we have noticed a major difference in grades between those students who read the minimum necessary to pass, perhaps not much more than the recommended readings, and those who pursue their interests with tenacity and curiosity.

We are often asked by our students how many books or articles they should read and reference for an assignment. This is a question similar to the proverbial 'How long is a piece of string?' We cannot provide you with an answer, but the question itself tells us that these students are still thinking of an MBA in mechanistic terms, rather than viewing it as a journey of exploration.

So, having avoided that question (justifiably we think!), what advice *can* we give you about reading?

Tips for effective literature searches and reading

Get to know your library

Ensure that you attend an induction to your business school or university library. If this is not offered on a formal basis, ask for a tour round so that you understand the layout and how to use the catalogues. Even if (and especially if) you are embarking on any form of distance-learning, getting familiar with your library will be highly relevant, as you will need to know what facilities the library can offer you through the internet. These days most good libraries have online

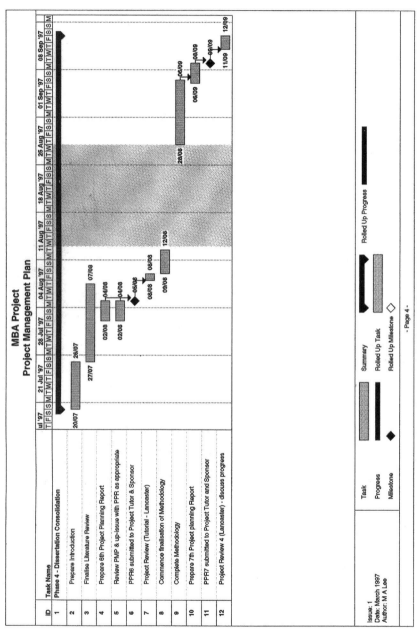

Figure 2.1 Example of a Gantt chart (reproduced with kind permission of Mark Lee)

access to a variety of databases containing all the major business, management and organization journals that are going to be indispensable to you. As a student on an MBA programme, you should be able to gain access to these, some of which may only be accessible from the library itself, but increasingly it is possible to gain access via the internet if you are armed with the appropriate password.

If you are studying at a distance from your library, find out what facilities they have set up for distant students. If they are unable to post books and articles to you, find the nearest good library to you and, if necessary, ask your business school for a letter of introduction. Many university libraries are happy to allow students from other universities to use their facilities, although they may charge a fee.

Tips for excellence!

Get to know your subject librarians. Many people make the mistake of assuming that librarians are people who stamp books and put them out on shelves! This could not be further from the truth. Librarianship is a highly specialized profession and, if you learn to use your librarians' knowledge and expertise effectively, they will become invaluable to you as you begin to refine your searches. Get to know the librarians who specialize in management and business and find out how much support they can offer you when you are conducting literature searches. Many of them are likely to have worked with countless MBA students over the years and they may be able to assist you in becoming more sophisticated in your literature searches, thereby finding more interesting and more relevant articles on the subjects you are researching.

Decide what to read

You will quickly notice that there is a limited number of hours in the day for reading, but an unlimited pool of books and articles, all of which seem both interesting and vital for your studies. So how are you going to select the most appropriate literature to read?

Firstly, listen carefully to your lecturers and tutors and note down any references they give you in class. Those of you studying at a distance will probably find your tutors are particularly helpful in this respect, as they will recognize that you will need extra support compared with those who have face-to-face access to them.

If possible, talk to past students of the programme to find out what texts they found helpful. This can often be an excellent indicator and sometimes more useful than course reading lists!

Some students think it is enough just to read the key recommended textbooks. Often they read these in their entirety to the exclusion of other texts. This not a good strategy, however, as you will only be able to write one-dimensional essays, containing a single view of the world, or if your textbook provides a summary of relevant research, you will only be able to cite a secondary (and not necessarily accurate) commentary rather than gaining an understanding from the authors themselves. An MBA programme is designed to help you to understand business and management questions from many angles. To do this, you will need to read diverse views of the same subject and, comparing and critiquing these, begin to form some views yourself. You will only learn to do this if you have found a wide enough set of views and approaches and ideally read these in their original forms instead of relying on 'set texts' alone.

In summary then, to decide on the most appropriate reading:

- First, follow up recommendations from your tutors.
- Go to your textbooks by all means, but follow up the authors cited in them in their original forms.
- Use the library catalogue to do searches by subject and authors.
- Use the journal databases to do searches by keywords. You can search by author, subject(s), or journal.
- Use online bookshops such as Amazon or BOL. These allow you to conduct searches by subject and author and often contain useful book reviews. They are also a quick way to obtain the books you need.

Tips for excellence!

The books and articles you start reading will all contain references to other work. Notice the references that crop up frequently – these may be important. Highlight the references to books, articles or authors that sound relevant to your study and follow these up where possible. Don't be put off if they are not immediately available; most libraries have an ordering service and your persistence may pay off by giving you an original angle on a topic which could enrich the quality of your argument. If you are unsure of the value of a book or article, or if you get stuck, ask your tutor or librarian for assistance as soon as possible.

Learn to search the library databases efficiently

There is a wealth of data now available through the internet, with most libraries subscribing to databases containing those academic journals which are relevant to management topics. These are too extensive to list here, but it is important to discover which are likely to contain articles relevant to your assignments. Doing searches requires patience and creativity. If, for example, you are interested in the impact of organizational culture on performance and key in the words 'organization' and 'culture' you are likely to be overwhelmed with 'hits'. Refining this search to achieve a manageable number of articles may take time. Try, for example, adding the dimensions 'organizational ideology' and 'symbolism' and you might now find yourself with a manageable number but whose content is not entirely relevant to your research question. By adding a further dimension, e.g. 'organizational performance', and refining the search with additional words such as 'management', 'productivity', 'climate', 'control', etc., and by trying different combinations, you will soon find you become adept at finding the most relevant and useful articles for your coursework or research question.

Get started early

Students often make the mistake of waiting until the last minute to get started and then discover that all the relevant books have been borrowed by other students. All libraries experience this to some extent, as despite holding multiple copies of core texts, the sudden demand for these at peak times is inevitable. If you start your search early, you will allow yourself time to order books and articles that are not immediately available on the shelves.

Effective reading

If you haven't studied at this intensity before you may be under the illusion that you are going to be able to read the books that have been recommended to you from page one to the end. Unfortunately, you will soon discover that, with a few important exceptions, this is not going to be possible. Adapting your reading style to meet the pressure you are under to cover a lot of theoretical ground within tight timescales is one of the hardest lessons that many students have to learn.

We have found a number of useful techniques that should help you to speed up your reading and avoid many wasted hours.

Skim reading

This technique involves scanning pages quickly to gain a feel for the content. This can be very helpful when faced with an article or chapter where you are not sure whether its content is likely to be relevant to your search.

Scanning contents pages

This technique is particularly important when you are reading for a purpose. Most of your MBA reading will be purposeful and you will have a particular question or theme in mind as you read. Scanning contents pages will help you to focus on the chapter(s) that might be relevant to your needs and interests. If you read these chapters and find that they are useful, you can always read the rest of the book later if there is time.

Reading introductory chapters

Reading introductory chapters can be equally invaluable for selecting chapters to focus on. Many books, particularly edited collections, contain a brief synopsis of each chapter in the Introduction.

Searching indexes for key words

Searching indexes is a technique that we find particularly helpful. Using keywords in the same way that you would when conducting a database search and scanning indexes for words relevant to your topic can often lead you straight to the most salient part of the book for your needs. Try different associated words until you strike lucky and, if there are no matches, then the chances are this book is not going to meet your needs. If the paragraphs you find using this technique appear to be relevant, go back to the start of the chapter in which they appear, so you can read the ideas in their wider context. If there is time, you can then read further into the book, but again prioritizing relevant chapters.

2.4 Recording your findings and taking effective notes

Although you may believe that your memory will serve you well and that you will remember the location and author of that important quotation, memory is very fallible when engaged in studies. It is possible to find the perfect quotation to illustrate your point, forget to make a note of where you found it and then waste hours searching through all the books and articles you have read trying to find it again when you come to write up your assignment. Believe us – we have done it and we have met many students who have duplicated their efforts in this way. Setting up an effective database or even a card index system for recording your reading can save many wasted hours later on. You will need a system that enables you to record the following pieces of information:

- author
- the title of the book or article
- the date of publication
- the publisher
- the page numbers from which you have quoted
- the quotations that you have used in full
- your notes on the article or chapter.

There are a number of computerized packages available for this purpose, but you can easily set one up with minimum technological assistance if you prefer.

Many students have found that they have benefited from using a daily logbook in which they have recorded their insights as they progressed though the programme. We recommend that you consider adopting the use of a logbook as a useful addition to your database. Your database will enable you to search your reading by author or topic, whereas your logbook will enable you to record your insights chronologically and therefore enable you to trace and reflect on the development of your thinking over the course of the programme. It is better if you choose bound notebooks for this, as loose-leafed pads have a habit of coming apart and pages falling out.

Many MBA programmes will ask you to reflect on your learning during the course. A logbook will enable you to capture your insights as they occur to you and you will be able to find them later when you need them. For example, we may see connections between texts, or disagreements, we may find ourselves surprised, irritated, disagreeing, or sympathizing with a text. The logbook is a useful place to note down these thoughts, as you are unlikely to remember them later!

If you make a point of taking the logbook with you to lectures and tutorials, as well as having it by your side when you are reading or working on an assignment, you will find it an important *aide-mémoire* when you come to doing examinations or your dissertation, when you might need to draw once more on the material you have covered.

Here are two more tips for recording your thoughts and keeping your data together:

- **Mind maps** are very popular with students for linking related ideas together. We have seen them successfully used in lectures for note-taking, and also in assignment planning for analysing the relationships across the literature, and for structuring the flow of an argument. As an example, we have included a mind map which Sharon produced when an MBA student herself, doing a module on organizational culture. This provided her with a useful prompt when she came to write her assignment on culture and also at exam time! See Figure 2.2 below. (For more information on mind maps, *see* Buzan and Buzan, 1996.)
- **Dictating machines** have been found by some MBA students to be invaluable, particularly by insomniacs and those who spend long periods in the car when it is not feasible to use a logbook.

We have also met students who express their ideas brilliantly when they speak, but lose this ability when they write. These machines have helped them a great deal, enabling them to articulate their ideas first before writing them down.

2.5 Planning for effective exam revision

We have devoted Chapter 12 entirely to the subject of exams, as we know this is one of the biggest hurdles in the MBA calendar and one that fills most people with trepidation. The point of including a paragraph on exam revision at this stage, however, is to say that the exam revision need not mean a last-minute panic. If your data recording and note-taking systems are well organized and up to date, you should not find it difficult to assemble your notes into subject areas and to include the main frameworks, theories and key authors from which you can revise. Remember when you set your systems up that aiding exam revision will be one of their purposes. (And remember that you do not want to start assembling your notes at that late stage.)

2.6 Summary

In this chapter we have focused on planning ahead for the challenge of the MBA by setting up your life to support your studies rather than hindering them – a problem that so many MBA students encounter as a result of poor preparation. The advice covered in this chapter on rebalancing your life and planning your studies should give you a head start, as well as a feeling of being in control.

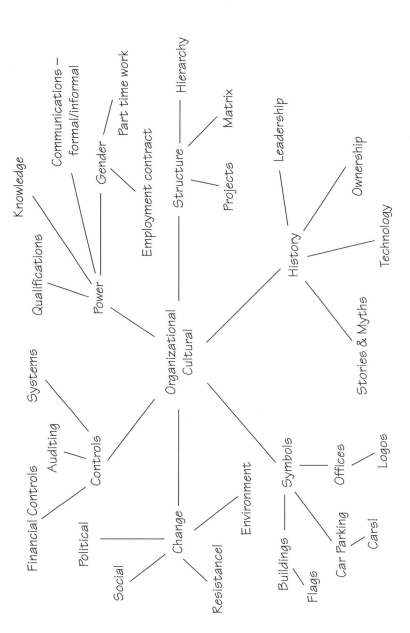

Figure 2.2 Sharon Turnbull's 'mind map'

Sharing knowledge and being supervised/tutored

3.5 Ten tips for sharing knowledge and being supervised

3.6 Summary

'Multi-cultural team projects were, for me, the most valuable part of the MBA – we were given very short timescales in which to complete the task set and this forced rapid learning of one another's cultural norms. This led to an understanding of factors that came into play later, when dealing in international business.'

Paul, Cambridge University

Introduction

Many people embarking on an MBA imagine themselves sitting for hours in a study, bedroom, dining room or library surrounded by books, burning the midnight oil, and making endless cups of coffee to meet ever tighter deadlines for coursework. Often this image comes from previous encounters with undergraduate study where the emphasis can often be on digesting large volumes of knowledge in the shortest possible time. Fortunately this is only a small part of the story of doing an MBA. Most people find the more participative style of MBA programmes much more fulfilling.

Although reading is fundamental to MBA success, much of your learning will also come from the involvement you have with your classmates and tutors on the programme. Whether you are studying on a full-time programme, meeting your classmates on a daily basis, studying intermittently on a part-time or modular MBA, or online on a distance-learning programme, it will be important that you make use of every opportunity to learn from other students. There is no doubt that students who make the most of all the learning opportunities available to them, including learning from the experience of others, generally achieve much higher grades than those who endeavour to do the whole course by focusing on textbooks and lectures alone. In other words, the old cliché 'What you get out directly relates to what you put in' is highly applicable to studying for an MBA.

We have noticed over our years of teaching on MBA programmes that some students learn quickly how to make effective use of their tutors and advisers, managing them well and tuning into their expertise and knowledge. Others, however, prefer to ' go it alone', often missing out on valuable opportunities for direction and discussion which might make their private study more efficient.

Purpose of this chapter

- To offer some ideas about what your tutors and advisers might be expecting from you (and what they don't expect!).
- To offer some guidance about making the best use of the tutoring available to you.
- To discuss how you might find ways to share your ideas and experience with other members of your class. Remember, most students will bring considerable experience with them to the programme, and missing the opportunity to tap into this important organizational and functional experience would be a considerable waste of a valuable resource.
- To offer guidelines for setting up and running a learning set, even if your course does not formally require you to do so.

3.1 What tutors/faculty advisers expect from their students

Your tutor or adviser can be one of your best sources of support or guidance throughout the programme if you can learn to use him or her effectively. Some MBA programmes allocate a personal tutor to each student for the duration of the programme; others offer tutorial support on a subject-specific basis. Whichever approach your course takes on this, it is certain that your relationship with your tutor(s) will be an important part of achieving success in your MBA.

Some students feel that to ask for advice or guidance is a sign of weakness, and prefer to work alone, often with disastrous results. These students do not contact their tutors in between attending lectures and submitting their assignments, preferring to be totally self-reliant. Be careful – this strategy rarely works! Not only is their learning diminished by this lack of involvement, but their grades often reflect this. Some people find making the transition from senior manager to student quite difficult at first, as it requires adopting a new identity, but remember that your tutors are used to helping people make this transition, so try not to resist their assistance!

At the other extreme, some students want their MBA to be 'spoon-fed' to them, looking to their tutors to do all the hard work for them, and to distil theory into manageable chunks that can be easily digested. This strategy is also likely to fail. In adopting such passive behaviours, these students are demonstrating that they are *not* taking responsibility for their own learning, a universal requirement of all good MBA programmes.

So what do your tutors expect from you? In summary, we suggest that your tutors will want you to strike a balance between these two extremes by:

- taking responsibility for directing your own studies
- acting on their advice
- actively and systematically engaging with practical and library research
- thoughtfully considering the application of these ideas to your practice as a manager
- demonstrating that you are taking responsibility for your own learning by pursuing and exploring all avenues of knowledge available to you.

Your tutors and faculty advisers will all differ in style, availability, interests and expertise, just as your colleagues on the programme will each bring different experience sets to the class. Your tutors are there to guide you on a very wide range of topics. The most commonly-asked questions by students of their tutors are about:

- the direction of their reading
- specific references
- the focus of their assignment research
- how to structure their writing
- how to choose a topic for an assignment
- how to collect their data
- how their coursework is progressing
- how they can improve their marks.

Do your best to establish an open, honest and trusting relationship with your tutors. They are there to help you to address all of the above questions and many more. They will provide important continuity throughout the whole programme and will appreciate it if they feel that you are using their expertise wisely and thoughtfully.

3.2 The difficulties and benefits of working in learning sets/tutorial groups (including online groups)

Working in learning sets or tutorial groups often requires a considerable adjustment in your mindset compared with other forms of learning, particularly for students who have been used to working individually and competitively in previous courses. Whilst MBAs do vary in the extent to which they encourage competition between students, what they have in common is the requirement that students learn to collaborate in order to achieve success. Arguably, this is an important skill for managers, and one which you have probably already learned in your managerial role; yet it is one which students are often reluctant to exercise within the boundaries of academic study.

The reason for this reluctance seems to have been built on four myths that we would like to dispel.

● **Myth 1** If I give away my best ideas to others, they may do better than me!
  In practice, the ideas that you formulate when working on your assignments are most meaningful to you and have invariably emerged from a combination of your own personal experiences, your reflections on the MBA lectures, and the reading which you have since been conducting. Therefore, they have an element of uniqueness that makes plagiarism quite unlikely. We would suggest from our experience that the risk of one of your colleagues copying your ideas is much less great than the potential benefits that can be derived from them helping you to develop and solidify your ideas during a group discussion.

- **Myth 2** I might inadvertently copy others' ideas if we discuss these during the preparation of our assignments and coursework. Whilst the ideas of others *are* likely to trigger some good ideas for your own work, and add to your own thinking, it is most unlikely that you will be tempted to 'copy' them, as each of you is likely to be taking your own unique and individual approach to your assignment work, even when the topics you are pursuing are the same or similar. Don't be afraid of sharing ideas. You will no doubt already be convinced of the amazing results that can be derived from brainstorming activities in the managerial world. The same creativity can be generated in your MBA studies if you are willing to collaborate with others.

- **Myth 3** I can do much better if I keep my thoughts to myself and remain self-sufficient.
  This belief suggests that despite any synergies and overlaps in the research paths that each of you are following, you can all produce better results working independently. Our experience has shown otherwise. When students take this view, there is little added value to be gained from learning sets. Groups that overcome this block invariably find that there is much to be shared in working together, and this approach can reduce individual workloads, as well as improving the clarity of ideas, and therefore the coursework of all of the group members.

- **Myth 4** I need to like and respect all members of my group in order to perform successfully.
  Sometimes students are fortunate in that all members of the group get on well. However, this is not always the case, and you may find that you are required to work with people whom you find difficult or do not like. There is never an easy answer to this situation, but in our experience group members can overcome these tensions if they are willing to work at it. The key is to focus on the task, communicate as openly as possible and remember that you all have the same end goal of extending your learning and achieving high grades. Even if the group members are not those whom you would choose as friends, try to look for their best assets, particularly in relation to your MBA.

3.3 How can learning sets work effectively on an MBA programme?

Your learning set may be pre-assigned by the course tutor, or you may be asked to choose the colleagues with whom you wish to study. Either way, it will be important to find ways to work together efficiently and successfully. The guidelines in this section apply equally to online learning sets, as do the points made above and indeed following these steps should help you to avoid many of the pitfalls that some of our own students have experienced in the early stages of set formation.

Understand your purpose

Learning sets can be used to help their membership to think through assignments or dissertations; to share resources when researching a topic, particularly when library searches are required; to provide feedback on coursework; for case study analysis; and even on some courses, when this is permitted, to undertake case examinations. Some sets form voluntarily to provide support for their members. Others are a compulsory and integral part of the MBA programme, designed to support an action learning philosophy and to assist students to apply theory to practice (*see* Chapter 9). There are many legitimate and useful purposes of learning sets and tutorial groups.

Make a contract

For a set to work effectively, each member must feel that the entire membership is committed to contributing equally to the group's learning and shares common assumptions about the purpose of the set. If some members contribute little but seem to be taking a lot, then trust will break down and the group will fragment. It is important, therefore, for you all to agree early on what you are prepared to contribute, the roles you will each play, the degree of sharing you would like to engage in, the time commitment you are able to give and how often you will meet prior to getting started.

For example, are you each willing to share important articles that you find by circulating these to other group members? Are you willing for others to read and comment on outlines or drafts of your coursework? Is it acceptable to contact each other if you are stuck at a point in your learning and need help in interpreting some course material?

The level of commitment you wish to agree upon is likely to vary, depending on the nature and style of your MBA, the stage you are at in your programme, your geographical proximity and many other factors. You may want to start off with minimal commitment and increase this as the trust between you develops and the need increases with the pressures of the course. It will be important to review this contract regularly as the programme progresses and your relationships develop.

Provided that your level of collaboration remains within the rules of your course,[1] you may use the set in whatever way assists you best, provided that you all agree.

If your tutor is involved in your learning set, it is also worthwhile ensuring that his or her level of involvement is also clear. Avoid making assumptions about any of the above that might result in acrimonious feelings later.

Understand each others' strengths

Whatever the purpose of your learning set, you will undoubtedly, at some stage, need to divide up your resources to make best use of the cumulative time available to you. Sets often omit an obvious step that they would rarely overlook in their managerial roles, that is to ascertain the specific strengths and expertise of each group member and divide tasks accordingly. Often this means playing to these strengths, particularly when the group is under time pressure to deliver results. Sometimes, however, for the purposes of learning and when the risks are less great, this may mean that group members try out new areas of expertise in order to learn new skills. So,

---

[1] It is very important that you check on this, particularly if you are setting up an informal learning set. Most MBA courses, for example, bar you from joint authorship of coursework.

for example, the accountant in your group may not always wish to take all the financial roles in the group and may sometimes prefer to work on human resource or marketing issues and vice versa.

Continually monitor the effectiveness of the group

From time to time it will be important to check whether the learning set is delivering to the individual group members what it was set up to deliver. Is it saving time through sharing responsibility for library searches, for example? Are you all finding it a useful environment to discuss your study-related and assignment-based problems and issues? Are you taking away good ideas from the meetings? Does each member of the group feel that they are getting out more than they are contributing? Is the set supportive of all its members and is it using each member's talents and experience effectively, as well as providing a forum for the development of new skills?

Take actions to deal with any problems identified

Don't despair if your set does not work perfectly straight away. Clearly, learning to work effectively in a learning set on an MBA programme is an iterative process, during which many of the questions asked above will need to be addressed more than once as each group experiments with different ways to work together and you start to accommodate each other's styles. There is no formulaic prescription to making a learning set work, except to be patient and remember that groups take time to perform successfully. Issues that you may need to confront early on might be if one member of the group seems to be excessively dominant in seeking to force their ideas onto the group, believing themselves to have the intellectual 'high ground'; or, by contrast, if one member does not seem to be contributing as much as the others. It is never easy to confront these issues, but our advice is to tackle them early. You will need to be sensitive towards the individual(s) concerned, but direct enough for them to understand that they need to change their behaviour. In the case of an over-dominant member, they may need your support in order to change. In the case of a weaker member, try to find out what is preventing them from making a full contribution, for example, whether it might be due to short-term personal problems and, if necessary, be creative in finding a more suitable role for them in the group.

3.4 Working across cultures

It is important at this stage for us to say a few words about working across cultures, because on many MBA programmes you will find yourself working with people from a wide range of national and cultural backgrounds. Space does not allow us to introduce here the theories and research appertaining to cultural differences. However, our experience has taught us that even in the act of setting up tutorial groups or learning sets, there is often a wide range of assumptions being made which may be hidden from the members for some time. Expectations and beliefs about gender, leadership, authority, age and etiquette may all vary in different cultures and inevitably lead to very different understandings and behaviours amongst group members. These can lead to misunderstandings unless they are brought to the surface and addressed. It is important, therefore, that as a group you find a way to understand and work with these differences early on. Try to find ways to discuss issues as they arise and, when differences in behaviours start to become obvious, ask your group members to try to articulate the assumptions which are driving their behaviour as far as they can. Openness to new ways of seeing the world and a desire by all members to understand the different cultural assumptions in the group will not only alleviate potential problems, but will open up a rich and important learning experience during your time on the MBA. We strongly suggest, therefore, that you take every opportunity to work with colleagues from other cultures, and, where you have a choice, that you avoid staying with those of the same nationality, even though at times this will be tempting!

The real-life examples below illustrate the value of taking cultural differences seriously.

Two real-life examples: Working across cultures

Example 3.1

A multi-cultural MBA learning set had come together for the first time and were asked to undertake an outdoor task which required them to build a bridge and transport all team members across a divide within a ten-minute interval. Soon an American and a German member of the team were each voicing different views on how the bridge might be constructed, battling with each other for their voices to be heard and canvassing opinions from the other team members in support of their ideas. A British and a French member actively engaged in the discussion, adding their views, while a Danish member started to sort out the equipment available and to experiment practically with the two designs. At the end of ten minutes, with the task still not accomplished, their tutor asked them to review the process they had been through. The Japanese member, who had quietly observed the débâcle, sat silently in a corner of the room until he was confronted angrily by the German member.

'Why are you so silent? Why did you not get involved? Don't you have a view?'

'I am a civil engineer,' he answered. 'I have the answer to the problem here on a piece of paper.'

'Why didn't you tell us!' shouted the German member, frustrated and perplexed at the apparently stubborn behaviour of her Japanese colleague.

Quietly and calmly he answered: 'Nobody asked me for my opinion. In Japan it would be rude to express views the way you two did without being asked for them. Now that you are asking me, I can tell you that there is a much better way to build that bridge than either of your ideas.'

At that point the tutor intervened to draw out some important learning points about cross-cultural team working.

Example 3.2

A group of Chinese students in an MBA class were feeling very frustrated by the low marks they had been awarded for their first assignments. They had worked round the clock to submit their first essay on time, had used books and articles extensively, checked their English and had felt that they fully understood the topic of the essay. By contrast, they noticed that some of their European colleagues who appeared to have spent less time than them and, indeed, were boasting about going clubbing the night before the submission was due, had achieved higher marks. The system seemed incomprehensible to them and they could not understand why the tutor was not happy with their work.

Feeling that the matter needed to be resolved, they asked for an appointment with Professor Rich, requesting that he give them more detailed feedback on what they would need to do to improve their marks.

'You have spent too much time *describing* the theory, just repeating what is in the textbooks,' said Professor Rich. 'There is no *critique* in your work; you have not even referenced the sources you have used, and you seem to have accepted the ideas of all the leading writers on this topic without question. Although it was clear that you had all done a lot of work, I could not give you higher marks because your assignments all suffered from the same failing – that they were uncritical and incompletely referenced.'

The Chinese students looked puzzled but nodded, thanked the professor, and without another word left the room. Professor Rich, feeling a bit uneasy about this exchange, pondered over the conversation most of the day. It was the first time so many students from China had been on the MBA programme and he was unfamiliar with their education system. Feeling slightly worried about their passive response to his explanation, he decided to invite the most vocal member of the group back to his office later that day.

'Tell me, ' he asked Roy, 'were you surprised by my feedback this morning?' 'Oh yes, we were, Professor,' he replied, then slowly went on to explain.

'In China we are not encouraged to criticize those in authority. Our studies are about gaining knowledge and demonstrating that we understand. We have never before been asked to criticize our masters, or the great intellectuals. We are very surprised that you ask us to do this. We are also surprised that you ask us to cite references for all the books we have used. This has not been expected of us previously on our courses in China – and it was certainly not our intention to plagiarize. We were a little shocked that you might have thought it was our intention to deliberately conceal our sources.'

Professor Rich looked thoughtful for a moment.

'Why did you not say something to me about this problem sooner – or at least this morning when you were in my office?' he asked.

'It would have been impolite to contradict you, Sir,' responded Roy. 'You did not ask for our comments. Now that we know what you require, we will work hard to learn how to deliver it for you.'

3.5 Ten tips for sharing knowledge and being supervised

Tip 1 Attend all your lectures, seminars, tutorial meetings and online discussions punctually

Where requests have been made for you to prepare for these by prior reading or case preparation, ensure that you allow yourself enough time to do this. Even if others don't do this, don't assume that it does not matter. Setting aside enough time to undertake careful preparation will greatly enhance the value of these meetings to you and others.

> **Tips for excellence!**
>
> Prepare thoroughly for your meetings. Your tutorials will always be more productive and time efficient if you are prepared. Bring a list of questions with you that have cropped up during the course of your study. It may sometimes be appropriate to e-mail these in advance to your tutor and tutorial group members in order to give others time to consider them and so be in a position to offer the best quality advice at the meeting.

Tip 2 Take an active involvement, not only in your own work, but also in the work of others

As indicated earlier, much of your own learning will come from discussions with others about the problems and issues they are dealing with. If you only appear to be alert when your own work is being discussed, your colleagues and tutor will notice this and may respond accordingly when it is your turn to seek guidance. We have noticed a remarkable correlation between active participation and a high level of contributions in set meetings and overall success in the programme.

Tip 3 Keep your tutor and learning set informed

If you absolutely have to miss any of the above meetings, let your tutor and set know in advance and give the reasons why. If your tutors and colleagues know you are experiencing problems they will be better able to support you. Students who disappear without letting others know why are likely to find little support or sympathy upon their return, however serious the nature of the problem. It may sound obvious, but this courtesy is often forgotten – a half minute e-mail or phone call can make all the difference to your relationships!

Tip 4 Be honest!

Remember that your tutors have probably been around for a while and have certainly been students too at some stage (often, like ourselves, on an MBA programme). They will see right through attempts to bluff your way out of trouble and will have heard most excuses for missed deadlines. Examples ranging from 'My secretary has left the country, leaving no forwarding address and has taken my assignment with her' (this really was used!) to the frequently used 'My hard drive blew up and I forgot to take a back up,' are not uncommon! We could write a book on excuses for missed assignments, and probably invented some of them ourselves, but our advice is *don't do it!*

If possible, don't miss course deadlines – it will only create a build-up of problems for you as the course progresses. However, if this is unavoidable, inform your tutor, giving a clear reason, as soon as possible.

Tip 5 Don't be afraid to ask for help

This leads us to our next important point. It is connected to the point above on being honest. We know that a late or missed assignment can be caused by numerous circumstances, for example, work pressures or a personal life crisis, but each circumstance demands a different response. Do ask for help if you feel that others may be able to assist. If you let others know when you are encountering problems, your colleagues and your tutors may think of creative ways to help you. Most people will experience some problems at some stage of their MBA programme. Fortunately most of these are minor and can be resolved, but in our experience it is those students who try to persevere alone who are most likely to struggle to complete the programme. (See Chapter 4 for more detailed advice on getting through the difficult times.)

Tip 6 Make appointments

If you need to see or speak to your tutor outside of scheduled meetings, try to make an appointment in advance. Remember that tutors have demands on their time and are often themselves involved in other programmes or conducting research to tight deadlines. They are more likely to give you their full attention if they have scheduled your meeting in their diary, than if you turn up at their door unannounced, however supportive or sympathetic they would like to be!

Tip 7 Listen to your tutor's advice and feedback

You need not always act on your tutor's advice and there will be times when you might differ in your view and choose to follow a different approach. However, it is irritating for a tutor who has invested a considerable amount of time offering guidance to find that this has apparently been ignored. If your tutor offers you advice, it is important that you consider the suggestions he or she has made, remembering that despite the wealth of experience you may have in *your* professional field, your tutor is the expert when it comes to success in an MBA programme. They want you to think for yourself, so will not feel slighted if you take an alternative perspective from their own, *provided* that you demonstrate the stages of your thinking and that you have not neglected their suggestions through laziness, complacency or disrespect!

You will receive feedback from your tutors on your coursework as you progress through the programme. This will be vital for your learning, so it is essential that you fully understand what is being said to you. If the feedback you receive is unclear to you, it is important that you ask your tutor for a further explanation, as you will need to build on this learning as you progress into the next part of the programme. Don't wait until your examination is looming before realizing that you did not understand what was being said to you about your coursework.

Tip 8 Give your tutor feedback on how things are going

Remember that communication is never only one way. As an MBA student, you are not simply a passive recipient of ideas, support and guidance. Your tutor relies on you for feedback on your learning and the learning of the group as a whole. Every class and every student is different, with different needs, learning styles and interests. Consequently, an open relationship in which you feel able to talk to your tutors about the programme and your needs in a constructive way will help to ensure that the programme design and content meet your needs as closely as possible. If you have been sponsored for the MBA, do make sure that you find opportunities for feedback to your employer about your progress on the programme.

Tip 9 Discuss your work regularly with your tutor and other members of faculty and tap into their expertise

Think of the faculty who teach you as a resource to assist you with your learning project. Be open to their challenges to your thinking when you get caught in (metaphorical) culs-de-sac. Ask them to share their own knowledge and research experience with you and to advise you of work by other scholars which might assist you as you progress in your studies. Engage with their critical and constructive feedback on your work, so that they can see that you are interested in stretching yourself and exploring ideas beyond the core curriculum of the programme.

Tip 10 Keep an open mind at all times and allow the diversity of the group to provide a rich learning experience

Although it is sometimes a challenge to understand how fellow students from different cultural backgrounds perceive the world and to understand why they behave in different ways in the classroom and in learning sets, do remember that the learning you gain from being part of a diverse group will be an invaluable part of your MBA, and will be even more invaluable in today's global business world. Remember that this rich learning environment is a rare opportunity – so make the most of it!

3.6 Summary

The key messages contained in this chapter are summarized below.

● Collaborate
  Set up as many opportunities as you can to share your ideas with your fellow students. Make the most of learning sets and be clear about how you want them to work.
● Communicate
  Use your tutors as a valuable resource. Don't be too proud to elicit their assistance if you get stuck, but don't expect to be 'spoon-fed' by them. Remember you are 'reading for' an MBA and they will expect you to do so!
● Empathize
  Be sensitive to cultural differences. A genuine interest in others' perspectives will enrich your learning many times over. Don't make assumptions about others' perspectives, ask about them, learn from them and, where appropriate, adapt your behaviours to accommodate others.

Getting through the difficult times

*'If you see the MBA as an answer to the problem of
what to do with your life, you will be disappointed –
the MBA cannot do that for you, because those
answers must come from inside yourself. But if it's an
answer to the problem of wanting to change your
career trajectory, the course may well help. In the
same vein, an MBA is a great way to collect lots of
answers – the questions that match them might or
might not exist, but you can have fun finding out!'*

Jonathan, City University Business School

Introduction

This chapter considers how personal and emotional issues can affect
MBA performance. In every MBA class it is inevitable that some students will experience a 'low period' during the programme. This
may be due to factors directly associated with the MBA programme.
Students may find themselves the 'non-conformist' member of a
group, or worry about being in a minority. Problems can also arise
due to the intrusion of 'life' events such as divorce, relocation or ill
health, which might make concentrating on academic work seem difficult, or even pointless at the time they occur.

Purpose of this chapter

Of course, this chapter cannot solve the problems for you, but the
authors hope that in writing it we may be able to prepare readers for
the 'lows' when they come. The chapter gives examples of the kind
of problems faced by students and gives some practical suggestions
about how to cope with these challenges in the context of MBA
study. It ends with some advice on how to deal with conflict or
grievances – which we hope you will never need but which may
prove valuable if things become really difficult. This chapter will
deal with specifics such as:

- issues for women MBA students
- doing an MBA in a country that is not your own
- personal problems
- issues for part-time and distance-learning students
- being the non-conformist member of a group
- dealing with conflict or grievances.

4.1 Issues for women MBA students

Many MBA programmes have a higher proportion of male students than female students. Even where numbers are equal, managing MBA study can be difficult for women because MBAs reflect the still male-dominated management world and if you are a female student, you may find yourself increasingly aware of the existence of the 'glass ceiling'. Although equal opportunities legislation has been established in many countries, women are still less likely to make it to the 'top' than their male counterparts. Some women find this becomes more of a stark reality when they begin an MBA because this coincides with the stage when they hope to leave middle management and make the transition to senior executive. At this point, it becomes obvious that while opportunities for professional women at the middle levels have improved, many institutions (such as governments, banks and even universities!) are still managed at the highest levels by men. This can seem disheartening for women struggling to gain their MBA in the knowledge that the qualification will not level the playing field.

In addition, women managers (or MBA students) can feel discouraged because they face criticism for adopting attitudes and behaviour traditionally associated with men. In her book *Challenging Women*, Su Maddock (1999:114) suggests this is due to long-standing social practices which 'presume women to be subservient, less competent and domestic'. What might be seen as laudable drive and ambition in a male manager can be regarded as confrontational and aggressive in a woman: 'Men who argue are regarded as rational whereas women are regarded as disagreeable' (Phillips and Pugh, 1994:118). Thus, it may feel as though students or staff in your own learning set can accept strongly expressed opinions from male group members, but are uncomfortable if women behave in the same manner. Clearly, neither this book nor your MBA is going to change the world. However, it is worth reminding yourself that an MBA will not only provide you with an excellent business qualification, but will also enhance your confidence, so gaining your MBA can only advance your position *vis-à-vis* the rest of the world. The following may be helpful advice to women facing the kind of issues outlined above, enabling you to complete your MBA programme to the best of your ability.

Helpful advice for women students

Some women do make it to the top

Remember that although opportunities in the management world are more limited for women than for men, opportunities do exist (this can be easy to forget if you are working on case studies where the executive personnel are all male). Search out some specific role models and read about them, or if possible, even meet them. How did they get where they are? What do they advise?

Make your point

Whether talking to other students or staff, don't be put off from making your point, but try to control the pitch of your voice. Keep both the level and tone low, but firm. This can help you fend off accusations of being 'irrational' or 'emotional' if you wish to disagree with something that has been said. Try to back your arguments with evidence. For example, a suggestion that working mothers are not committed to their careers could be parried by references to successful women who have combined family and career.

Enter discussions eye to eye

One of the authors received this 'tip' from a male friend who uses a wheelchair, and has found that it works well. If you are working with a group of male students (or managers) and you are noticeably smaller than they are, try to get everybody seated before beginning discussions or negotiations. This will avoid the possibility of anybody literally 'talking down' to you, using their height to intimidate you. It will also prevent a situation where you cannot join in the conversation because – perhaps quite unintentionally – the rest of the group is literally talking above your head. You will find it much easier to make your point without raising your voice if you can make eye contact with the rest of your group. (Note that this tip can also be useful if you are very tall and worry that your height could be intimidating.)

Establish networks

Get to know other women who are doing MBAs. Find out how they cope and share experiences. If possible, get to know some of the female teaching staff in your faculty, who might be able to offer support and advice if you have worries in relation to your MBA programme.

Reactions to your gender

Remember that although they may not realize it, others (including other women) will react to your gender. Are people threatened by you? Or conversely, are they afraid of challenging you because you are female and they fear an emotional upset? Often, people are reluctant to make feelings such as these explicit, but if you yourself can try and understand the reason for others' behaviour, this can help you to decide what action to take.

Change can be a long time coming...

Finally, it is important to bear in mind that although the need for change may seem urgent, revolution can be a long time in coming. Women in the UK have had the same voting rights as men since the 1920s, but there are still some private associations (some golf clubs might be one example) which do not accept women as full members and refuse them the right to vote. Irritating though it may seem, a little patience can go a long way and, if you are 'ahead of your time,' it might be better to focus on your reading and coursework while you wait for others to catch up, rather than spending valuable study time trying to whip them into action.

A real-life example: Gender

Example 4.1

Maria Standish was tiny, pretty and clever. She was a dealer working for a bank which funded her MBA at a top university. Maria worried that some male staff in the faculty disliked her assertive behaviour, which was not what they expected from someone with her delicate physical appearance. Maria felt that this did not matter in terms of her written work, which was marked anonymously. However, she was worried about her formal MBA presentation, when she feared she might be criticized for an approach which was considered more acceptable when exhibited by male students. Maria took this up with her (male) faculty adviser. She found him more open than she had expected and sympathetic to her concerns. He raised the matter with his colleagues who agreed that Maria's panel would be chaired by a female member of staff. Maria felt confident that the faculty would be fair to her because she had made her concerns explicit and these had been taken seriously. At her presentation, she made her points quietly but firmly, backing them up with evidence, and was awarded good marks.

4.2 Doing an MBA in a country that is not

your own

Ethnic and cultural issues

If you are from a different cultural or ethnic background from most of the MBA students on your programme, this can be daunting. Most universities try hard to help new students settle in. Nevertheless, if you are a long way from home, coping with unfamiliar foods, currencies and weather conditions, you can find yourself feeling excluded from what is going on. Often this is not intentional on the part of fellow students, but occurs as a result of thoughtless or poorly informed behaviour. Furthermore, if you are an overseas student, you may be obliged to work in a language which is not your own. Some suggestions are given below about how you might cope with this, as well as a real-life example of a student who survived being in the minority when studying overseas.

The language of MBAs

It might seem unfair, but the majority of MBA programmes (even those offered in Nordic and Far-Eastern countries) are taught in English and written assignments are required in English. You will be obliged to give presentations, and to work with other students, in English. As teaching staff, the authors of this book are aware that some students, whose first language is not English, face major difficulties with their MBA course due to a poor grasp of the language. Vague understanding of lectures and barely acceptable writing skills are not enough to ensure success. If not tackled, the problem of poor English does not get better but builds up to a crescendo of misery when students are trying to write up their final dissertations. Doing well on any Masters degree programme is difficult when you can manage the language well. If you are struggling to communicate, taking your MBA could certainly be an unhappy experience and may even be doomed to failure.

What should you do if you are already on an MBA programme, but are struggling with your English? Don't ignore the problem! If you have been allocated a personal tutor, or faculty adviser, speak to her or him as a matter of urgency, to see what can be done. Some universities have departments which specialize in helping overseas students improve their English language. If not, they may be able to recommend a private tutor. If you are really struggling, it would be worth taking a year out while you become more familiar with English. It is better to tackle the problem 'head on' rather than risk failure because your language skills are insufficient for you to understand what is being taught.

Homesickness

Feelings of isolation are a common problem among people studying overseas. They can affect mature students just as much as younger undergraduates. As a starting point, your university will arrange special events to welcome new students. Try and make the effort to go to these and find out whether there are any groups or associations of people from your own background that you can join. Don't forget to look outside the university because (especially if you are based in a city) you may find like-minded people within the local community. Discovering an outside interest – perhaps learning something new, or volunteering for something where your skills will be valued – can also help get you through. If you can afford it, try to get home at least once during your programme, even if only for a short time. Don't be 'brave' when family members ask how you are getting on. Explain how you are feeling and (especially if a visit home is out of the question) see whether a friend or relative might be able to come and visit you. In this electronic age you will at least be able to maintain regular contact with home via e-mail. Do use this and encourage friends and family to send you regular news.

Real-life example: Homesickness

Example 4.2

Venita Mukarji was an Indian student, taking her MBA in the West. Her language and study skills were good and she achieved high grades in her coursework. However, although a mature student, Venita had never lived away from her parents' home before and found the concept of living alone on campus daunting. This meant that she became very 'low' and found things difficult to cope with. Venita felt that other students lacked sympathy with the problems she was experiencing and showed no understanding of her extreme homesickness, which only served to make her feel more isolated. Venita coped with this problem by persuading two family members to visit her on separate occasions during her MBA programme. She also found a female staff member who was prepared to spend time with her, just sitting and talking. In addition, Venita looked outside the university for company and joined a local amateur dramatics group, which, to her surprise, she really enjoyed. It remained the case that Venita found studying abroad a difficult and sometimes unhappy experience. However, by thinking of ways in which she could combat her homesickness, Venita survived the programme and obtained a good result.

4.3 Personal problems

There are times in anyone's career when life 'gets in the way' and inevitably, some students will find themselves facing personal crises or health problems at the same time as studying. As discussed in Chapter 2, it is vital to take action sooner rather than later. Speak to your faculty adviser or course director as early as possible and warn them of the problem. If your difficulty is health-related, get a doctor's note. Your programme director may need this in the future if she or he is to plead extenuating circumstances on your behalf at an examination board. Depending on what the issue is, seek the faculty's advice on what you should do, and be realistic about what is manageable. If you feel you can cope with the problem and wish to continue with your MBA, ask for permission to review the situation again soon, and don't make promises about deadlines that you can't keep. Check whether allowances are usually made for people in your situation – and if they exist, don't be too proud to accept what is offered. If you feel it would be wiser to suspend your studies, *make sure that you and the faculty have agreed the terms and conditions for you rejoining the MBA programme*. This will be of value to both yourself and your programme director, because it will avoid misunderstandings when you return to your course.

Some questions you will need to ask are as follows:

- Will all the work you have done so far count towards your MBA, or will some assignments need to be re-done?
- Will your suspension of studies be time-limited by your faculty and, if so, how long are you allowed?
- Will you be able to re-join the MBA programme without extra cost, or will there be additional fees?
- Will you be charged additional fees if you are unable to re-join the MBA programme?

Make sure you have all this written down, ideally in a letter from the faculty to you, but if not, in a letter which you send to your programme director confirming details and inviting her/him to contact you with any queries.

What if you have not told anyone in your faculty that you are struggling with personal problems, but have nevertheless got into a mess with your work? As above, don't delay, but speak to a staff member as soon as possible. If you can show that your work has been of a consistently high quality, which has fallen only in relation to the problem, you will probably find staff on your programme sympathetic and willing to help out. If, on the other hand, you already have a record for poor quality work that is usually handed in late, it will be important to be open and explain why this has been the case, so that you can win their support for any further concessions you are going to need. They may even have noticed that something was wrong and were wondering what this is. The important thing to remember is that they cannot help you if they are unaware that you have a problem!

4.4 Part-time and distance-learning students

The above advice is equally relevant whether you are studying full- or part-time. Part-time and distance-learning students, however, can often suffer particular problems in relation to 'getting behind'. As a part-time student, you are likely to be holding down a demanding job and you may have family responsibilities. Your MBA programme could be stretched over three years or more. It will inevitably mean working at weekends and, unless your company is very supportive, formal study weeks at your university may have to be taken as annual leave, which can take a toll on personal relationships. It is unsurprising that this group of students often finds it difficult to keep up with their studies when faced with other, unexpected pressures in their lives. Ironically, however, this is also the group of students most likely to be shy of admitting that they are in a mess with their work. This is for a number of reasons. In the first place, part-timers are not spending as much time on campus as full-time students, so they may not know teaching staff well enough to feel comfortable about discussing personal difficulties. Conversely, MBA teaching staff may not know part-time and distance-learning students as well as their full-time counterparts and are less likely to notice when there is a problem. Thus, it is possible for part-time students to get quite behind before anyone from their faculty realizes that something has gone wrong. Distance-learning students' main form of communication is via e-mail and they may not feel happy about writing their problems down and sending them over the net.

Often, part-time and distance-learning students may be more familiar with a course secretary or administrator than with the academic staff. If so, try contacting whoever you know best to explain your difficulties. It's a start and once you have taken the first step towards dealing with the problem, you are on the way to resolving it.

4.5 The non-conformist member of a group

What if you find that you do not 'fit in' with your MBA group (either your own learning set, or with the MBA class as a whole)? All groups develop their own ways of working and ground rules, even if these are not formally stated. Most groups have at least one 'non-conformist' member and it can be disconcerting to find that you are the only person who does not, or cannot, harmonize with everyone else. Perhaps the first thing to bear in mind is that being the 'odd one out' need not mean that you are 'wrong', even if everyone else seems to think so. Remember that Emmeline Pankhurst and Martin Luther King were non-conformists! As the group 'dissenter' you may be an agent for change – someone who sees things differently and wants to widen boundaries. A non-conformist group member often notices, and makes explicit, things that others fail to observe or choose to ignore (such as matters relating to power, gender or ethnicity). This can make others feel uncomfortable, or threatened – especially if the observation is accompanied by a request for change.

If you find yourself the 'non-conformist' in your MBA class, you need to work out how to survive your MBA course and complete it, without compromising your ideals. One of the ways of doing this is to resist the temptation to clash with others 'head on'. Most importantly, try to avoid being enticed into conflicts that do not concern you, because people are often tempted to ascribe blame to non-conformists. Furthermore, since you are someone who sees things 'differently' from others, it might help if you think about how others see things. You may feel *you* know where boundaries should lie – but where does everyone else think they are? Understanding this might go a long way toward helping you turn conflict into a valuable chance for reflection for yourself, if not for the rest of the group. For example, you may believe that modern management structures should be flat. A group member who has made their way to the top of a traditional hierarchy could find this very threatening. Having understood this, you might approach the situation in a different way. Rather than just arguing your viewpoint, you could try getting other group members to think about the pros and cons of both styles of management, which will probably gain you more support for your views than if you simply continue to argue your corner.

You must also be careful about challenging teaching staff in public. In general, an enquiring attitude from a student will be welcomed because it offers the class a chance for debate and discussion. However, lecturers are only human and if they find themselves continually tested to the limit, they may lose patience (as might the rest of the class, who may be ready to move on to a new subject). If you do question the opinions expressed by your lecturer or faculty adviser in front of others, make sure that you are thoughtful and courteous in your approach, that you allow the staff member an opportunity to respond and that you listen to their reply.

Another way in which you can express your views is through your written work. This must be done with caution. Above all, avoid producing a polemic, which broadcasts your thoughts without any back up. If, like Jack Haines, in the real-life example given below, you are able to fulfil the course requirements, you may then go on to make your point – as long as you have found substantive evidence in the literature to back you up. You could, for example, argue the rationale for more flexible working hours for parents – but *only* if you make a strong business case, using data and arguments from the literature to support your line of thought. *A political diatribe will not gain you good marks.*

Real-life example: Making your point

Example 4.3

Jack Haines was a US citizen working for a British company who gained his MBA in the UK. He found himself the only American on a part-time programme with mostly European students. Many of the business examples used on Jack's MBA focused not on Europe, but on American companies. Jack found this difficult because many of his classmates and even some of the European professors made sweeping statements about the USA which he considered to be inaccurate and out-dated; for example, the implication that 'all' American managers were white, middle-class, wealthy and male. Jack's own experience did not accord with these generalizations, but he did not wish to be in the position of criticizing class teachers in public. Jack decided to argue his points through his assignments. As well as considering the course texts in his written work, Jack read widely and backed his views with firm evidence. There were still moments in class when Jack had to grit his teeth and refrain from arguing, which was not easy for him. However, he was able to make his point through his written work without compromising his grades, which meant he could relax and enjoy the programme, without feeling that he had failed to raise issues that were important to him.

4.6 Dealing with conflict or grievances

What if you believe that you are being unfairly treated by other students, or staff on your MBA programme? What if you feel that your department has made insufficient allowances for your personal problems? For a small percentage of students experiencing such difficulties, it may be appropriate to seek help from your faculty. It is not easy to define exactly when this would be, but as a guide it is unreasonable for your peers – or anyone else – to assess you on factors that relate to you personally, rather than your MBA performance. Your ethnic and cultural background, gender, sexual orientation and any disability you may have should *not* be relevant to the way you are treated by others, or the opportunities you are afforded on the programme. You might also wish to talk to your department if you wish to question the grade awarded to a piece of work or claim extenuating circumstances for an assignment that has gone badly.

Whatever your concern may be, the way in which you tackle it can make a big difference not only to the end result, but also to your own well-being. When the going gets tough (especially since the advent of e-mail), students can be tempted to rush to their PCs, shooting off written complaints without giving this a second thought. More often than not, this is something they will later regret. Complaining should be the last resort, not the first line of defence, because it is time consuming and stressful with no guarantee of success. If you do make a formal complaint, you have to be prepared for the possibility that it may not be upheld, or that it could make you unpopular with other students or staff on your programme. Of course, there are instances (often reported in the press) where long-standing and discriminatory traditions are overturned and individuals are compensated. However, these successes are rare (which is why they appear in the press), and they often occur only when the individual has left the institution concerned – probably without a degree, or job or whatever they were there for in the first place. If you are unhappy with some aspect of your MBA but wish to remain on your degree programme, it is better to try first to resolve the problem yourself before embarking on the process of making a formal complaint. Your main aim should be to negotiate a rapid, amicable compromise which is acceptable to all concerned and which will enable you to continue with your studies. If you feel you really cannot sort out the problem, you may decide to complain formally and your faculty will undoubtedly have a procedure which you can consult. However, the staff on your programme are likely to want you to be motivated and successful on your programme. Once you approach them, you may well find them willing to help. They need to understand what the problem is and you, in turn, need to know exactly what you seek from them. Before making a formal complaint, therefore, it may be helpful if you think through the following suggestions.

Approaching your faculty

'Before you begin, consider, and when you have considered, act.'
Cicero

What are your motives for wishing to complain?

Think seriously about why you are feeling aggrieved. For example, if you are wishing to query a grade, is this because others seem to have done better than you have? Or have you done better in previous assignments, meaning that you were expecting a higher mark than the one received? Before complaining, ask for some feedback from your faculty and consider this carefully.

What do you want your faculty to actually *do* about this?

What do you hope to achieve by approaching your faculty? It is important to decide what you want your institution/faculty to *do* in order to put things right. For example, in a situation where you feel that students in your learning set are excluding you, do you want your faculty to change the rules about learning sets? Do you want the faculty to tackle your group on your behalf? Or do you simply wish to be moved to another group?

Is the problem affecting your grades?

Are you so unhappy that your ability to concentrate and succeed on your programme has been compromised, or is the problem just generally irritating, i.e. do you have to deal with the matter now, while you are in the middle of your course? Or could you put it to the back of your mind, leaving yourself with the option of writing a letter to the faculty after you have completed your degree, if you still feel strongly?

Have you tried to resolve the matter yourself informally before approaching your faculty formally?

If not, it is worth having a real try to deal with the matter yourself first. Weigh up in your own mind whether your problem has arisen due to thoughtlessness and insensitivity on someone else's part (in which case you could try and talk them round) or whether they are being unkind/unreasonable on purpose.

Are you the only person affected by this?

Are there others who might feel the same as you do, or are you the 'lone voice', or non-conformist member of your group (see pages 64–6). Perhaps people do not intend to be unfair, but just see things differently from the way you do.

Does your faculty, or institution, have a formal policy relating to the issue concerned?

If so, this is likely to be published somewhere. Look on the website or in the student handbook, so that you gain a clear picture about your institution's approaches to matters such as equal opportunities, discrimination, sickness absence, carer leave, etc. Do students have the right to ask for their work to be sent to an external examiner, or is this always up to the relevant department?

Is there anyone who can advise or, if necessary, represent you?

There may be a staff member within the faculty who is sympathetic to your problem and will stand up for you. It is easy to forget (especially if you are an older graduate student) that students' unions and associations exist for you, as well as for undergraduates. Even if you choose to tackle the matter yourself, a student association might be able to advise you about how best to approach the situation. If the first person you speak to is unsympathetic, *don't give up* – try someone else!

Remain focused

When you are unhappy, and in the middle of trying to sort out a grievance, it is easy to get 'sidetracked' and forget what you are trying to achieve. Try and be clear about where your 'boundaries' lie. Resist getting sucked into a process which is more complicated than it need be. For example, if you want your work to be re-appraised by an external examiner, you should not get involved in trying to change internal marking procedures. Likewise, if you seek a personal apology from someone who made a rude remark, you do not need to take responsibility for getting your faculty to change its policies. Consider in advance whether or not you might be prepared to accept something 'less' than what you have asked for. Would reappraisal of your work by a member of staff from outside your department suffice, or do you need to stand firm and insist on it being sent to an expert outside your institution? Would a verbal apology from the person who made the remark be acceptable, or do you seek a written expression of regret?

Remain calm

This is far easier said than done. However, if you can be polite and firm throughout, you are less likely to be accused of 'over-reacting', or 'getting things out of perspective'. You are also more likely to retain sight of what you are aiming to achieve. One way of remaining calm is to avoid discussing the issue with other students unless you are with someone you know well. In this way, you will avoid it becoming the only thing you are able to talk or think about. This will enable you to concentrate on your studies while discussions about your grievance are ongoing. Above all, remember that it is in the interests of your faculty for MBA students to do well – so it is in their interest, as well as yours, to seek a speedy and appropriate resolution to your problem.

Querying the grade on your assignment

One common cause of disagreement between students and faculty advisers or other staff is the grade awarded to an assignment. Students often feel unhappy if they believe they have worked hard on something which receives a low grade, or if their grade drops below that received for previous pieces of work and they do not understand why. Before taking action on this issue, look carefully at the written comments you have received to see if they explain why the grade is low. If not, go and talk (calmly and politely) to your faculty adviser and ask how the marks have been awarded. If you are still unhappy, you will probably find that your MBA programme has a facility whereby students may request re-marking by a second examiner who may be external to your institution. Our suggestion is that it is only worth taking this option if you are a borderline case for a distinction or a fail. If you do decide to press for this, it is important to remember that the word of the external examiner is final – and that, like share prices, the marks of an assignment sent for external or second marking can go down as well as up!

4.7 Summary

This chapter has discussed the problems of trying to do well on an MBA programme during times of stress and unhappiness. Suggestions are given to help those who find themselves in a minority, and the issues faced by overseas students, women and 'non-conformist' group members are discussed. Advice on how to deal with conflict or grievances is also given. The general tips which summarize our approach are listed below.

- If possible, try not to manage your problem alone – don't be ashamed or nervous of seeking help from others, including fellow students or faculty staff.
- Keep lines of communication open between yourself and your faculty – they cannot help you if they are not aware of your problem.
- Try to see if there is any possibility that you could resolve the problem by seeking help informally from faculty staff or others before making a formal complaint.
- Be clear about what it is you want and whether you are prepared to accept a compromise when you bring the matter to the attention of your faculty.
- Try to remain calm and measured when discussing the problem – then no-one can accuse you of losing your temper or being irrational.

Making the most of classroom learning

'Never be afraid to ask the professor if you don't understand something. It doesn't mean you're stupid, or you wouldn't have been admitted in the first place. If anything, it shows you're thinking about the subject.'

Daisy, Simon Fraser University

Introduction

Whether you are in the classroom for a year as a full-time MBA student, or only for a week during the summer as a distance-learning student, it is vital that you make the most of your time in the classroom environment. You may worry about being taught as part of a large group, or about being asked to speak in a discursive environment. Be reassured, however, that being in an MBA classroom can be a positive experience and the advice below is aimed at helping you to get the most out of what is on offer.

Purpose of this chapter

- To address your concerns and to enhance your effectiveness in the classroom.
- To bring you up to date with current pedagogical approaches to learning and prepare you for what you might encounter in the classroom during your MBA.
- To consider the cultural differences which influence interaction in the classroom and, in particular, how these may affect your learning if you are studying your MBA in a culture which is not your own.
- To ask you to think about the classroom style that you are familiar with and to consider how you might adapt your behaviours to the style of the programme on which you have enrolled.

5.1 Cultural differences

Many MBA students come from countries where classroom participation is rare (e.g. in China and other Asian countries) and where most classroom activity takes the form of lectures. The expectation is often that students should listen carefully and take notes in order to make sense of the material later, outside of the classroom. This sense-making activity is generally an individual activity performed through private study. The relationship between students and tutors in these cultures may be distant, and students are encouraged to be recipients of knowledge passed to them by their professors. The model in these countries is based on a clear distinction between the tutor as expert and the student as learner.

In MBA programmes in Europe and the USA, by contrast, the classroom model is usually very different from the one described above, with a high level of participation being required from all students. The relationship between students and tutors in these programmes is much closer and students are encouraged to share their own knowledge and experience in the classroom, as well as to work in groups on joint problem-solving activities. The role of tutor or adviser as expert is diluted in this model, with the tutor's role as a facilitator of learning being emphasized much more strongly.

During our many years of teaching on and directing MBA programmes we have frequently met students from overseas who have told us that adapting to the totally different learning environment and culture in the classroom was the most demanding aspect of adjusting to their new environment, and that this was much more challenging to them than reading and writing in the English language. The challenges that they often report include knowing when they are expected to speak, how to formulate their ideas quickly enough to be able to interject in a fast-moving debate, how to behave in small group work, how to ask for clarification, when to challenge their tutors and their peers, and how to cooperate and collaborate effectively with others. Induction courses and support systems often deal with linguistic issues, but rarely do they adequately cover the pedagogical differences to be encountered in the classroom as a result of different academic traditions. This chapter endeavours to point out some of the main aspects of classroom learning and style in European and American classrooms that students from other cultures (or indeed those who are returning to education after a long gap) may encounter.

5.2 The lecture

The term 'lecture' is still widely used in academic circles, but the meaning of the term has evolved considerably as new approaches to learning have been developed in higher education. At its core, a lecture is still likely to involve a structured presentation from an expert in the field. However, it is possible that you will also be encouraged to interrupt throughout the lecture with your questions or comments. Most lecturers will let you know when they would like you to do this. This is a serious request, because if you have a question to pose, it is quite likely that your colleagues will also learn a great deal from the response and discussion that your question might generate. As well as welcoming your questions and comments, it is likely that the lecturer will pose questions that will encourage you to make connections between the topic being discussed and your own experience or reading. These questions may sometimes be addressed to an individual, but more often are offered to the class as a whole, which may then lead to a wider debate between students in the class. These are rarely asked as rhetorical questions and most lecturers do genuinely want a reply!

Think about your lectures as offering an introduction to a topic. They can only 'scratch the surface' in the time available and subsequently it will be up to you to follow up the ideas with your own reading and assignments. So do ensure that you have asked for (and recorded in the appropriate place) references for any recommended articles or books which will allow you to follow up on the ideas after the lecture. Try to find out early how the topic is going to be assessed, whether there is an assignment attached to it and, if so, what the assignment question will be. Understanding how the lecture fits into the bigger picture of your studies will help you to focus on the most relevant aspects and get the most learning from it.

Ten golden rules for success in a lecture

Below are our ten golden rules to help you get more out of your lectures.

- **Rule 1** Arrive early and find the most suitable position in the room to enhance your learning. If you are a little shy about making a contribution, you might find that sitting at the front makes it easier for you to speak as the size of the room seems less daunting.
- **Rule 2** Don't be too afraid to ask questions. It is likely that others are wondering the same thing. In most MBA classrooms, posing your questions openly to the lecturer (or to the class) will be considered as a helpful intervention and will often provoke a useful discussion by raising points not yet covered, or by encouraging the lecturer to reframe a point for greater clarity.
- **Rule 3** Don't be afraid to make *relevant* comments about the material being delivered. It is quite likely that thoughts that occur to you about the application of the material to practice will also be helpful learning for your colleagues and will be welcomed by the lecturer.
- **Rule 4** Do try to ensure that you contribute frequently to these discussions. Don't undervalue your ideas. Even if they are not yet fully formulated, expressing them in the classroom will enable others to help you to develop them and your tutor to understand how well you are making sense of the material being covered. If you need a little time to formulate your point in English, explain this to the group. Don't wait until the moment passes – it is better that you ask for airtime at the moment it feels important and relevant, even if the class has to exercise patience as you search for the right words.
- **Rule 5** Don't worry if others put forward a counter argument that appears to disagree with your own. Different points of view may be equally valid and your colleagues may be making an excellent point, in which case you will probably learn from it. It may be that your own idea is more persuasive than theirs, but you need to reformulate it or find new evidence in support of it in order to persuade others. This will also help you to learn and to develop your ability to argue a case, a skill that will be important throughout the MBA programme.

- **Rule 6** Don't worry about speaking too much! Most students need to be encouraged to contribute more in class and it is rare to find participants who speak too much in class. However, do check your points for relevance – taking the discussion off on a tangent can be quite disruptive for the flow of the lecture.
- **Rule 7** Pay attention to the signals you are getting from the lecturer or the class. If your question or point is felt to be relevant, the lecturer will almost certainly ask you or other participants to develop it further in the classroom, and will be more than happy to devote time to it. If, however, they judge that it is more important to return to a more central theme, they will signal this to you, at which point it is better to save your point for another time, or perhaps for outside the classroom.
- **Rule 8** Don't be afraid to challenge – but do this sensitively! Most lecturers will be happy for you to challenge them provided that you do this constructively and courteously.
- **Rule 9** Be equally sensitive and constructive when challenging your fellow students. Whilst the implications of embarrassing a colleague may be less severe, it does not make good sense to alienate yourself from your classmates, however 'right' or clever you feel in doing so. Learn to argue tactfully. The strength and persuasiveness of your arguments are what will impress your class and tutors and this can be achieved without scoring points over others!
- **Rule 10** Listen carefully and take good notes! It is important that you balance your class contributions with active listening, not just to the lecturer but also to the contributions of your classmates. Don't assume that the only useful knowledge will come from the lecturer. Most MBA courses are full of participants with many years of managerial experience to offer to the class. Find ways to capture what is being said in a way that is comprehensible to you later. Many students develop their own shorthand, or draw mind maps to make thematic connections (*see* Chapter 2). Set up a system for filing your notes so that you can retrieve them easily later.

5.3 The case study

Since case studies are an important part of classroom learning, we have devoted Chapter 8 exclusively to this topic. Case studies may be used inside or outside the classroom and they are usually a group activity. Inside the classroom, they are often used as a consolidating activity, in other words, as a way of practically demonstrating aspects of a topic which might already have been introduced through a lecture. They are an excellent way to check understanding, as well as to illustrate theory in practice through 'real-life' examples.

If you have a tendency to be passive during lectures, the case study is an alternative way for you to become actively involved. It will often require division of labour within a small group, so this will enable you either to play to your strengths or to try out new roles within a relatively safe learning environment. Many students feel more immediately comfortable participating in case study activities as they usually require working in smaller groups and for some it may feel less daunting making a persuasive argument in this size of group than in the plenary sessions. If you feel intimidated or shy about making your points when in the full class, it is a good idea to practise arguing your points during case study preparation in smaller groups in order to build your confidence for speaking in the larger group. If your MBA course has a convention of 'cold calling' on students to open the case (as at Harvard), it is vital to ensure that you are always well prepared, especially if you are assessed on your contribution.

5.4 Simulations and role plays

Not as common as lectures and case studies but nevertheless excellent vehicles for learning are simulation events. These usually involve participating in simulating the activities of a company or organization, often with MBA participants playing the various roles of key players in the organization, making key business decisions in 'real time'. There are many variants on this theme; for example, it is common for tutors to play the role of a board of company directors facing serious problems and to invite the MBA participants to play the role of consultants who will undertake a diagnostic study of the organization with a view to making recommendations to the 'Board'.

Some MBA students find role playing a difficult experience, either feeling self-conscious about playing a part that is not their own, or perhaps finding it difficult to imagine being in the part of the organization or the job they have been assigned to. It is important to try to overcome these feelings in order to get the most out of the simulation event. The point of simulations is not simply to think rationally about the problems facing an organization and to propose solutions to these (as would be the case when undertaking a case study exercise). In a simulation you will actually be 'living through' the problems and, consequently, any decisions you make will have real repercussions for you to manage during the life of your simulated organization. You might, for example, be a managing director having to make some of your workforce redundant, or explain your policies to angry shareholders. Or you might find yourself playing the role of a finance director tasked with controlling the budget, and faced with an angry research and development director demanding extra funding for a new project. Role playing can be a highly emotive experience as you live out these scenarios. You can learn a great deal about managerial decision making, power and politics, strategy, etc., but you can only do this by fully and actively participating, in order to feel the intensity of time pressures, intractable problems, strategic dilemmas, etc. Offer to play roles with which you are unfamiliar and wish to learn more about. Not only will you experience the role, you will also learn more about the pressures that these managers face.

In our experience such events are difficult, costly and time-consuming to stage-manage, but the learning that can be gained from them can be enormous. Make the most of any opportunity you have to be involved. Simulations are now also being developed for the purposes of networked learning, so that distance- and e-learning students need not miss this opportunity to role play in cyberspace!

5.5 Group tasks/discussions

We have already said a great deal about group work in Chapter 3. We have included this topic here, however, to remind you that it is not only in learning sets or tutorial groups that you will have to work closely with others. Indeed, on many courses there is a substantial amount of group work which takes place in the classroom and which involves groups of any size tackling any task set by the tutor. These often involve applying models and theories to practice by drawing on the participants' own experience, or debating a series of complex questions. Frequently these small groups will be asked to feedback their findings to the plenary group.

In designing group work for MBA programmes, course directors are conscious that management is not a solitary activity and that students need to learn how to maximize the effectiveness of working in a team. In Chapter 3 we discuss the cultural issues of working in groups, as well as issues of gender, boundaries, support, power, etc. In the classroom these can often be magnified because time constraints for such exercises can often be tight, causing emotions to run high. Often the exercises set for classroom work have an element of competition built into them, adding to their intensity. In Chapter 3 we focus on learning sets as a mechanism set up to support your individual learning. Group work in the classroom differs in that it is the success of the group which is often at stake in these exercises, not that of the individuals within it. This means that frequently individuals are required to sacrifice their personal agendas for the greater good of the whole group. Group work in the classroom serves multiple purposes on the MBA course. Your consistent involvement and participation, commitment to overcoming difficulties of understanding, willingness to compromise, work towards common goals and to deal with difficult situations will form excellent grounding, not just for success in your MBA, but also in your post-MBA career. Our advice is to be persistent, patient and tolerant. Don't despair when you run into difficulties; tackle them directly, don't 'lose your voice' and, above all, don't opt out – to do so would be to miss a great deal of valuable (though often painful!) learning.

5.6 Summary

In this chapter we have reintroduced you to the classroom in the context of MBA programmes. We have discussed the most common forms of activity you will be asked to engage with in the classroom and made some suggestions about how to get the most out of your time there. Whilst the time you spend in the classroom as part of your MBA may only be a minor component of your overall learning, it can be a very important opportunity to seek out new ideas, interact with stimulating colleagues and ask the important questions you are not able to do when studying on your own. Our ten tips for success and enjoyment in the classroom are as follows.

- Try to find out early about expected practice and classroom 'norms' in the school where you are studying, but don't worry if others' behaviour in the classroom differs from yours because of their different cultural backgrounds. Remember that cultural diversity can add richness to the learning in the class.
- Prepare for your lectures as thoroughly as possible by undertaking in advance any suggested pre-reading.
- Think of the lecture as an introduction to new ideas, not as an end in itself.
- Follow up on interesting ideas as soon as possible after each lecture.
- Engage fully with all activities in class, especially those which push you outside your 'comfort zone'.
- Listen carefully to others and consider the views of your fellow students as actively as those of your lecturer.
- Interact as much as possible in class.
- Pay attention to feedback and signals from your group or tutor relating to your contributions and your classroom behaviour.
- Take good notes and file them in a place where you can easily retrieve them later.
- Speak up if you don't understand. Your colleagues will appreciate your honesty and may be experiencing similar difficulties themselves.

Learning at a distance – making the most of distance- and e-learning

6.5 How to manage the difficult times when studying at a distance

6.6 Summary

'Distance learning was an excellent experience as it gave flexibility to fit into other areas of life and the teaching materials were top quality. (In fact I still use them seven years after graduating.) The tutorials were helpful and fun, giving an opportunity for working with many different people. I know I found studying by distance-learning easier than a modular programme as it did give so much flexibility.'

Sarah, Open University

Introduction

An increasing number of MBA programmes contain some element of distance-learning and some are designed for the entire course to be taught in this way. Much of what we have said in earlier chapters of this book is, of course, also relevant to those of you studying at a distance. However, this chapter deals specifically with the issues which specifically relate to distance- and e-learning.

Purpose of this chapter

- To explore what is meant by distance and e-learning.
- To discuss the advantages and disadvantages of the virtual learning environment.
- To consider how studying at a distance impacts on sharing knowledge and being supervised.
- To consider what additional study skills you might need in order to be successful studying at a distance.
- To consider how to manage the difficult times when studying at a distance.

6.1 What is meant by distance-learning and e-learning ?

Distance-learning options have been available to MBA students for about twenty years. Even before the advent of new technologies, complete MBA programmes were available in paper form. The increasing sophistication of the virtual environment, however, has meant that in recent years much of the paper which formed the basis of distance-learning courses has been replaced by online documentation, making postal problems almost a thing of the past.

Although e-learning has been growing rapidly over the last few years, it is still a relatively novel term, and many business schools are still developing the technologies and expertise to enable them to deliver their courses online to the same standard as face-to-face and classroom learning. Its growing popularity is partly as a result of the increasing access to internet technology and partly due to students from overseas wishing to take advantage of a European or American business school education. It is also because many managers are reluctant to give up prestigious jobs to study for an MBA full-time, preferring instead to study for an MBA part-time, with maximum flexibility, whilst continuing with their full-time jobs.

There are, of course, many forms of distance-learning and e-learning, and in this chapter we can only touch on a few.

MBA by distance-learning

The content of these programmes is very similar to taught MBA programmes, the main difference being the mode in which the modules are taught. Many courses package each module in the form of a workbook (online or paper format). This includes readings, case studies, exercises, assignments, etc. Students work through these in their own time, submitting regular assignments to their tutors, who then mark these and return them with comments and marks. Deadlines and timescales for completion of these modules are much more flexible than on a conventional MBA. Often the order in which a student completes the modules is also flexible, and the maximum period for completion for the degree is usually much more generous than for a conventional MBA, as the design of the programme recognizes that the time that students have to devote to their studies will vary.

In their early forms, many distance-learning MBAs suffered from very low completion rates, with many students reporting feelings of isolation. As a result, distance-learning courses now include learning sets, 'summer schools' and tutorials, in which students can discuss the material and any difficulties they may have.

With the advent of web-based technology, many of these enhancements have now been incorporated into online learning environments, leading to the growth of networked learning as a form of distance-learning.

Networked learning MBAs

With technology evolving so fast, it is not possible to describe all aspects of these learning environments. However, most of them include bespoke conference sites containing course materials, links to relevant websites and library databases, online lectures available through digital video technology or live broadcast, discussion groups, and much more. The latest technology is even starting to introduce avatars[1] to the learning environment.

Many of the old problems of distance-learning have been overcome in networked learning environments. Communication between students and tutors flows more easily and, in theory, only the reliability of the technology and the competence of the users should act as a barrier to progress.

However, studying an MBA through an e-learning environment does present different challenges from studying in a face-to-face environment. The remaining sections of this chapter will address these challenges.

---

[1] An avatar is an icon representing yourself, which can attend virtual classes and conferences along with the avatars of other members of the group.

6.2 Managing the advantages and disadvantages of distance-learning

Whilst many of the issues faced by distance-learning students are similar to those studying on more conventional MBA courses, there are a number of important advantages and disadvantages, which will be experienced by those studying online. Awareness of these differences can go a long way towards helping you to sustain your motivation and interest and to find creative ways to 'stay the course'.

Advantages

More flexible time-management

With the increasing sophistication of information technology and networked learning environments, there are many advantages to studying at a distance. It usually means that you do not need to block out periods of time to attend classes and, as a consequence, you will experience less disruption to your professional life, as you can build your studying into your timetable and schedule it at times to suit yourself. This is particularly useful for those whose work patterns are irregular or who travel extensively.

Ease and flexibility of communication

Using web-based technology means that, provided others in the group go online regularly, there is no reason why you cannot be in almost daily contact with your MBA group. Successful online study environments will be buzzing with activity – with students discussing their ideas on the latest module, posting drafts of their work for comment, making suggestions about helpful books and articles and tutors facilitating these discussions as they would in a face-to-face environment. Measures are put in place to protect confidentiality. Through the use of different passwords which determine access, these discussions can either be restricted to subgroups or open to the whole MBA cohort. Time differences across the world are immaterial when learning is taking place in a networked environment, although real-time discussions can also be organized if the group wishes.

Effective use of tutorial time

E-learning means that you can ask for your tutor's help and feedback when you most need it, instead of being constrained to wait until the next tutorial meeting. Most tutors would prefer to engage in discussions with you when you are ready to do so, rather than meet you at a predetermined tutorial time, only to find that you are not really ready to discuss your work. Of course, this flexibility does have limits as your tutors are busy too, but generally, if you manage this well, it will allow you to make very effective use of their time.

Disadvantages

Sustaining motivation

For many students it can be much more difficult to motivate themselves when they do not have the discipline of regular face-to-face contact with their group and their tutors. In our experience, the first thing that happens when students fall behind or lose motivation in a networked learning programme is that they stop logging on to the conference site, making it impossible for their classmates or tutors to know what has happened or to help or support them. As time passes, it then becomes increasingly difficult to re-enter the discussion, and eventually some of these students give up on their studies, often citing lack of support or feelings of isolation as reasons for their withdrawal. It is much easier to avoid contact with your colleagues and tutors when you are studying online than it is when you have to attend regular classes and tutorials and this is a clear disadvantage. If you are studying on a conventional course, your tutors will quickly realize if you are experiencing problems and can then take rapid action to support you. It may take much longer for them to discover that you have a problem when you are studying in an e-learning environment, particularly if you are studying at your own pace and are not connected to a specific MBA cohort.

Isolation

A related challenge for e-MBA students is how to find ways to keep the connection between yourself and other students and to develop friendships. Without a real effort to maintain regular contact with others online, it is very easy to start to feel isolated. It is much easier for these feelings to go unnoticed online than it is in a face-to-face environment, where it is more difficult to hide any problems you are dealing with. Some students are looking for more than intellectual stimulation on their MBA. It might sound obvious, but if you seek social (face-to-face) contact with others, an online MBA is unlikely to be the right mode of study for you. If you are already enrolled on an online MBA and it doesn't fulfil your need to meet with others and share experiences, it might be worth transferring to a part-time programme with a residential component.

The internet is not the place for company secrets

One of the major disadvantages of learning online is the possible leaking of confidential information to large numbers of people, by accident. This is usually related to human error, rather than faults in the software. We have all observed the embarrassment experienced by colleagues who, intending to mail a personal message to one person, have accidentally pressed 'mail to all', thus ensuring that everyone in the organization (or wider) receives a copy. This is one risk of learning online and may be something you cannot control. Perhaps *you* are careful about what is sent to whom, but other students may be less stringent in their approach. Our advice is to exercise caution over what you send over the internet, which is never the place for company secrets or intimate confessions.

6.3 How does studying at a distance impact on sharing knowledge and being supervised?

Many of the issues that we discuss in Chapter 3 are the same ones that those studying at a distance will face, but clearly the opportunities for you to 'bond' socially with your group are greatly constrained if you are dispersed throughout the country or indeed the world. You will, therefore, need to work much harder at relationships than your peers on conventional programmes for whom making friendships will be quite easy.

Managing your tutor

In theory, studying online should make managing your tutor easy. If you send regular updates on your progress and details of any questions you may have, they should be able to deal with these at a time convenient for themselves. However, these same tutors may also be simultaneously running conventional courses and the students on these courses may be making counter-claims on their time. These students are there on the spot and able to make appointments to see them, whereas you may be many miles away. Whilst your tutors will want to give you the same amount of time as your peers on conventional programmes, you may nevertheless be competing for a finite amount of time, and it is therefore important that you have strategies for ensuring that your own work is dealt with promptly and does not slip down your tutor's list of priorities because you are not there in person.

Agreeing a date by when your tutor will feed back to you is a useful strategy. This will ensure that they plan their feedback to you and, if this does not arrive by the agreed date, you can legitimately remind them. Do bear in mind, though, that contracting in this way places obligations on you in the first place to meet your own deadlines, in order to let them have your work on or before the agreed date.

Telephoning your tutor to let him or her know that you have posted some work online may also be a good strategy if your tutor tends not to go online every day. This way you will not lose valuable time and you may also be able to reinforce your main points or questions in a telephone call. This way your work will be at the forefront of your tutor's mind and may well move up their list of priorities as a result. (Don't overdo this strategy or you may start to irritate your already very busy tutor!) Allow reasonable timescales for their responses or they may start to see you as unreasonable and you could inadvertently alienate yourself.

Building good relationships

Work hard at building good relationships with your programme director, administrator, secretary and school librarian – indeed, with all those whose support you may need during the course. Try to personalize your communications so that they remember who you are. Developing a relationship without having met a person is much more difficult than it is in face-to-face communication, but if those who can help you are already feeling predisposed to assist you, this will help a great deal. Remember that they are likely to be experiencing their own work and life pressures just as you are, so try to empathize with them when they are busy and they will be more likely to do everything possible to help when you do have an urgent problem that needs their assistance. If possible, visit your school at the beginning of the course to meet these key people, as it will mean that the online relationship will develop more quickly. Don't worry if you can't, however – there are many ways around it. Think about how you would behave if you were meeting them face-to-face for the first time: you would probably engage in some light conversation about your work and life prior to engaging in business. Why not try this online? You may be surprised how successfully it helps you break the ice, even without the benefit of a face-to-face meeting.

Studying in an online learning set

In order to provide support for online students, many MBA programmes organize online learning sets to encourage ongoing mutual support, both pastoral and academic. These will only work if all members make an effort and, whilst the same principles apply to building relationships in these groups as they do for face-to-face groups (*see also* Chapter 3), you will have to work harder at making these successful since you will not have the same opportunities to meet socially and break down any barriers between you.

Opportunities to air feelings are sometimes more constrained online than face-to-face. This need not be the case, however, and you are encouraged to be as open as you can with your fellow group members online. Bear in mind, however, that dealing with sensitive issues is not always easy in electronic communication as e-mails and website postings are unable to contain the nuances and tonality that are possible when holding a conversation. Indicating that you are being humorous, light-hearted, ironic, etc., is very difficult online and there are times when it might be better to speak to a person live, rather than risk misunderstanding. These times are rare, however, and provided that you exercise care, you should be able to operate as successfully online as you would face-to-face.

Discussing and contracting how you will work together early on in the life of a set will help you to avoid problems and misunderstandings later on. You will need to agree how often you will each go online, the level and type of feedback you will give each other, the amount of sharing of papers and references you will do, as well as finding a process for supporting one another during difficult times.

Our experience has shown us that a well-functioning online set can often be even more rewarding and successful than those that meet face-to-face. Many students report that the quality of the feedback, support and contributions from their peers is often better online than face-to-face, as this form of working allows the time and space for considered reflection that is not always the case in a tutorial meeting.

6.4 What additional study skills will you need in order to be successful studying at a distance?

Some students are fearful that their IT skills will not be good enough to enable them to communicate successfully online. In most cases this is an unnecessary fear, as these days most conferencing packages are designed to be extremely user-friendly, with clear icons to indicate the different pages available and a 'help' facility for those who need it.

You will, of course, need to learn to find your way round the internet, and use search engines and library databases to find articles and books, but these days even students on conventional MBA programmes need these skills too. If you are inexperienced in doing searches, start practising soon. There are books and courses that can help you with this, but most people find it relatively straightforward and prefer to learn the skill through hands-on practice!

It is also worth investing in the best infrastructure you can afford (for example, Broadband) to maximize efficiency in terms of quicker downloads. This will easily pay for itself in time saved.

6.5 How to manage the difficult times when studying at a distance

Most of the barriers to success in a distance-learning MBA have nothing to do with the acquisition of specific IT skills. The feedback we have gathered has shown, by contrast, that the main challenges to success in distance-learning are organization, motivation and relationship-building. It is, therefore, to the following three key criteria that we now turn:

- planning your studies
- sustaining your own interest
- building strong links with other students and your course tutors.

This section reinforces the messages in Chapter 4 on managing the difficult times. These messages are even more important for distance-learning students than for those on classroom-based programmes, and the importance of communicating regularly and clearly with other students in your cohort and with your course tutors is even greater if you are studying at a distance.

If you are encountering difficulties in your personal or work life, and falling behind with your studies, it is crucial that you advise your tutor immediately so that you can jointly decide how to address this problem. Whereas when teaching on a classroom-based programme, tutors can sometimes guess that something is going wrong in a student's life, on a distance-learning programme this becomes more difficult, particularly as one of the first signs of distress is often a failure to log on, followed by the complete disappearance of the student. Don't let this happen to you. The following tips are designed to help you to meet the three key challenges listed above and help you to get through the difficult times.

Planning your studies

- Always send copies of your work plans to your tutor for approval.
- Check with your tutor, colleagues and family that this plan is feasible, and build in breaks for holidays, family activities and busy periods at work.

- Each time that you revise your plans, send these to your tutor so that s/he knows when to expect work from you.
- Involve your work colleagues as much as possible in your plans, so that they know when you need space for your studies.
- If you fall behind with your plans, let your tutor and learning set know as soon as possible.
- Ask for help to get back on track.

Sustaining your own interest

- Discipline yourself to go online regularly. The latest messages from your set members should help to sustain your interest.
- Share interesting articles with others and try to stimulate debate online.
- Try to share your ideas with your friends and family. You may not have face-to-face contact with your fellow students, but you do have other people around you who may be interested in what you are learning.
- Don't just rely on course materials. Do additional searches and follow interesting lines of enquiry, particularly where you can imagine their application in your workplace.
- Try out some of your new ideas at work and set up discussions and workshops with colleagues in order to share your new learning.
- If you find your motivation waning, ask yourself why this is happening. Are you struggling to understand the relevance of the material? Are you studying a subject you are already familiar with? Are you finding difficulty finding interesting articles on the subject? Your tutor and learning set should be able to help you address any of these common problems. Try to explain how you feel so they can help you.
- If your mind has been taken off your studies as a result of personal problems, do let your tutor know about it so that they can support you through this time and help you to cope. Most tutors are genuinely sympathetic to those experiencing personal problems outside the programme.

Building strong links with other students and your course tutors

You may not be able to socialize face-to-face very often (if at all) if you are studying at a distance, but you can find other ways to build relationships with your fellow students. Some ideas are listed below.

- Try to build a relationship with your fellow students that goes beyond the needs of the programme.
- If opportunities arise to meet any of your group, take advantage of these.
- Share with the group some of the other aspects of your life that are important to you. This will enable them to understand the factors influencing your commitment levels to the programme. Find out about other members of the group too. You may find common interests beyond the MBA programme. If you are facing personal problems, try to share them. This will help the group to find the most appropriate way to support you.
- Keep the group abreast of developments at your workplace and find out about theirs. Some learning sets successfully use their group as a sounding board for business ideas and problems.
- If you need specific help from the group or tutor ask for it – don't wait for them to guess what you need.
- Try to build reviews of the group process and dynamics into your discussions. If relationships are not working, sensitively raise the issues early so that the group can address them before they escalate.

6.6 Summary

This chapter has highlighted the main differences in the experience of studying at a distance compared with that of studying on conventional programmes. We have introduced you to some of the many forms of distance-learning and discussed some of the e-learning variants of MBA programmes. Focusing on the advantages and disadvantages of distance-learning, we have suggested strategies for managing your relationships, time and commitment when studying at a distance. We have concluded that the greatest additional challenges facing distance-learning students are: sustaining your motivation, organizing your studies and building relationships. We have offered tips for each of these. This chapter should be read alongside the main chapters dealing with these issues, in particular, Chapters 2, 3 and 4.

In conclusion, we would remind our readers that although we have contrasted classroom-based programmes with distance- or e-learning programmes as though they were discrete entities, in reality the boundaries between the two are becoming increasingly blurred. Even on full-time and face-to-face modular programmes, we are seeing an increasing use of the networked learning environment to support classroom learning, in the same way that many distance-learning programmes offer residential blocks to their students once or twice a year. This blurring of the boundaries is, it seems, set to continue, and we expect that many students on 'conventional' programmes will find that they will benefit from reading this chapter if parts of their course are now available online.

7

Choosing electives

'Some people chose their electives on the basis of the style of the lecturers, the most amusing being the most popular, but without really thinking of their longer term career needs. I tried to focus on the content of the courses as my top priority, and was very glad, when I was promoted, that I had done so, as two of my chosen electives turned out to be very important in my new job.'

Marc, De Paul University, Chicago

Introduction

Most MBA programmes will ask you at some stage in the programme to choose between elective courses. The choices you make will be very important, not only to your success on the programme, but also for informing and supporting you in your future career. Without some carefully chosen criteria for making your selection, it would be easy to make decisions that you might later regret, and it is for this reason that we felt it important to devote a chapter to this topic.

In our experience, there are many unwise reasons for choosing an elective. Here are some examples:

- 'All my friends are doing it!'
- 'It's the only one I know anything about.'
- 'The lecturer was persuasive and asked us to sign up on the spot. We didn't want to disappoint her.'
- 'I left it too late so it was the only elective with any vacancies.'
- 'The lecturer's got a nice bottom.' (Actual words of a female student overheard by one of the authors!)

In this chapter we offer a structured approach that should help you to avoid the traps illustrated above. Making elective decisions is not easy as there are so many factors at stake. Many of you will still be feeling uncertain about what you want to do after your MBA, so making decisions without a complete understanding of your future needs may feel very uncomfortable.

Purpose of this chapter

- To outline a structured approach to eliminating those electives that will not serve you well on the programme and in the future, in order to select those that will serve your interests in as many ways as possible.
- To help you to avoid making any unwise choices and thus to maximize the use of your time and efforts on the programme.
- To offer you a process for reviewing the content of the electives on offer and for reviewing your needs and desires (current and future) in order to make an informed decision.

7.1 Gathering information

Electives can vary widely in content, style, level of challenge, breadth, depth, difficulty and commitment required. You will need to balance all of your current needs, paying attention to your personal circumstances and commitments, as well as to your success on the programme so far, with those long-term needs that you foresee beyond your MBA when you re-enter the labour market. This will not be easy. If you are currently struggling with the demands of the programme it may not be sensible to take on an elective that will add to this pressure and you may need to place your current well-being above your future needs. On the other hand, you may decide to forego a potentially fascinating elective in favour of one that appears to offer some important practical skills which you may offer to a future employer. Taking care over this process invariably reaps great rewards.

In the following step-by-step process we ask you to respond to three key questions, before rating each elective offered against the criteria you have set. The questions are:

1 What do I already know about the electives and what further information do I need?
2 What do I want from an elective?
3 What other factors are important to me in making my decision?

Please follow carefully the steps outlined below. By listing your criteria for each question and then weighting these, you should be in a good position to make some well-informed judgments about what is right for you.

Question 1 What do I already know about the electives and what further information do I need ?

Sometimes information about electives can be a bit piecemeal as often each lecturer is responsible for putting together details of their own course. This sometimes leads to differences in the quantity and quality of the information you have been given. So what do you need to know?

Below we have offered you an initial checklist, but suggest that you add your own criteria below ours.

- How long is the elective?
- How much contact time is involved?
- How much private study is involved?
- What is the nature of the assessed work?
- What is the content of the course?
- How demanding will it be?
- How much do I already know about the topic?
- Will I need to do any preparation for the course?
- Who is going to teach it?
- What do I know about their style and approach?
- When is the elective being run?
- Where is the elective being run?
- Is there a distance-learning option for this elective?
- Other criteria (*please list below*).

It is important before moving to the next stage that you obtain information to answer all of the questions listed above, as well as those you have added. If the published information does not give the details you need, don't hesitate to approach your tutor, course leader, or administrator to be sure you have all the facts available to you to make your decision.

Once you are satisfied that you have a clear response to each question above for each elective, you can move on to the next question. Try to avoid jumping to early decisions on the basis of the information you now have. Continuing to follow this structured approach will help you to consider all of your needs and to look beyond the immediate appeal of the elective course!

Question 2 What do I want from an elective?

We suggest that you list your criteria in the same way that you did for Question 1 above. We have started the list for you, but please add any other criteria that occur to you. Against each criteria rate its importance as follows:

5: Vital
4: Very important
3: Quite important
2: Not very important
1: Unimportant

- Will be useful in my future career   ____
- Will be mentally stimulating   ____
- Will be intrinsically interesting   ____
- Will introduce me to a new academic area   ____
- Will have clear practical applications   ____
- Will allow me to build on a subject I already understand   ____
- Is a leading-edge topic   ____
- Will offer me a new skill   ____
- Will offer me new theoretical frameworks   ____
- Other (*please list below*)   ____

Once you have rated each of these, do the same exercise for the next question using the same rating scale.

Question 3 What other factors are important to me in making my decision?

Once again we have started a list for you but do add your own criteria to this list, then rate the factors as you did above.

5: Vital
4: Very important
3: Quite important
2: Not very important
1: Unimportant

- The lecturer is interesting and delivers well   ____
- The lecturer is an expert in the field   ____
- I expect to do well in this topic   ____
- It will not make too many demands on my time   ____
- The timing of the course fits in with other commitments   ____
- It is offering a unique opportunity   ____
- It can be studied at a distance   ____
- It offers travel opportunities   ____
- It offers the opportunity to study with new people   ____
- Other (*please list below*)   ____

7.2 Applying your criteria to the electives

Now that you have completed your responses to the three questions outlined above, you are now in a much stronger position from which to consider each elective in turn.

Using the knowledge you gathered during your research into Question 1 – for each elective – consider the criteria you have listed in response to Question 2, 'What do I want from an elective?'

Place a tick against all of those factors which apply to this elective (to which you can answer 'yes'). In the example below, Jane Brown has ticked 6 out of the 10 possible criteria. Now count how many of the factors ticked have scored a 5, then a 4, then a 3 and so on.

For example, Jane Brown's result looks like this:

Elective 1: Managing Information

- Will be useful in my future career   5 ✓
- Will be mentally stimulating   3 ✓
- Will be intrinsically interesting   5
- Will introduce me to a new academic area   3 ✓
- Will have clear practical applications   1
- Will allow me to build on a subject I already understand   4 ✓
- Is a leading-edge topic   3 ✓
- Will offer me a new skill   5
- Will offer me new theoretical frameworks   2 ✓
- Other (*please list below*)   ____

There are six ticks, distributed as follows:

1 – 5	(5)
1 – 4	(4)
3 – 3s	(9)
1 – 2	(2)
0 – 1	(0)

Jane's total score for this elective will be 20.

For the next elective, Managing Change, by contrast, Jane has counted the following:

Four ticks, distributed as follows:

1 – 5	(5)
2 – 4s	(8)
0 – 3s	(0)
1 – 2	(2)
0 – 1	(0)

Her total score for this elective is 15, making this a less attractive option.

Once you have completed this exercise for all electives available to you, we suggest that you do the same exercise for Question 3, 'What other factors are important to me in making my decision?'

Once you have completed your scoring for both questions, you will now have two 'total' scores for each elective, i.e. one for the question: 'What do I want from an elective?' and one for 'What other factors are important to me?' Now add together these two scores. You should now be able to rank the electives according to how successfully each one meets your criteria, starting with the elective which scored the highest total number of points.

How vital is *vital*?

You will have noticed as you followed this process that you indicated in your list that a number of criteria were *vital*, i.e. you rated them as 5.

Now that you have a ranking for your electives, you should go back to those electives that you have rated as most closely matching your combined sets of criteria to ensure that none of your top electives contains criteria which you rated as vital yet remain unticked.

If we return now to our example of the elective Managing Information, Jane had indicated that 'Will offer me a new skill' was vital to her, yet despite its high overall score, this elective does not appear to meet this important criterion. Jane will now need to ask herself again whether this criterion is really vital, particularly since all other factors have scored highly for this elective. If it is, and if she believes that this elective cannot offer this possibility, she may decide to rule out the elective on this basis and proceed to the next highest scoring elective where all 5-rated factors have been awarded a tick. She may, on the other hand, downgrade this requirement, thereby deciding that offering a new skill is not after all vital. If this is the case, she may then decide not to rule the elective out on the basis of this single factor.

It is, of course, a matter of judgement how *vital* a vital factor is for you and whether you wish to compromise on one factor where all the other criteria are scoring highly. These final refinements to the process can be quite difficult and time-consuming, but can be very important to ensure that you give some thought to any anomalies, such as the one illustrated above.

7.3 Further tips

We believe that the process outlined above will lead you to a carefully thought through decision. We know, however, that in the final instance, and particularly in the event of a tie between electives, some of your more subjective criteria might come into play. We also know that you might like to adapt our template by adding further questions in order to meet your own needs.

The most important advice we offer in this chapter is that you should adopt a systematic approach to decision making and then follow it rigorously until you are certain that you have considered all your options.

Finally, we offer a few further tips to reinforce the need to be systematic and to offer a few further ways to locate the information you need in order to make your decision. These are listed below.

- The success of this exercise is entirely dependent on the quality of the information you have gathered about the elective, so don't be tempted to skip Question 1. Answering these questions thoroughly will enable you to make much better judgements about Questions 2 and 3.
- If in doubt about an elective, do seek out the course leader if you can and have a chat about its appropriateness for you. Go armed with a list of questions you would like to know to help you to make your decision. The Question 1 checklist might be a useful starting point.
- Consider who else you might ask about the electives. Previous students, if available to you, are an ideal source of vital information, particularly about the style and quality of the course. Your tutor may also be able to guide you, as s/he will have a good idea about your own strengths and interests on the course. Find out how long the elective has been running, and ask whether there are course ratings available from previous years in order to find out how past students have rated these electives.
- Start your research early so that you are not pressured into taking a course which does not really interest you.

- Don't be influenced by other group members' choices. Their criteria might be entirely different to your own. It is fine to discuss these, but remember that everyone's needs and interests are different.
- If it helps to do this exercise with a friend, ask them to play a facilitating rather than an advisory role. Only you can make the decision that is right for you.
- Think about the long term as well as the short term. You will need to balance your current needs for an interesting and stimulating course with future expected career needs. Sometimes you will need to prioritize one of these for the other and this will be one of the most difficult decisions to make, particularly if your future is still uncertain.
- Don't underestimate the importance of these decisions. Electives can be a vital and stimulating part of an MBA if chosen wisely, but can be a burden to you if chosen in haste. We are convinced that the time you invest in considering your options will be most worthwhile.

7.4 Summary

In this chapter we have argued that choosing electives is a vital aspect of your MBA programme and can be an important aspect of achieving success in your marks. We have offered a systematic approach for making these decisions that you may use as it stands or adapt to suit your own needs. We have strongly suggested that you invest some time in considering your choices in order to make the most appropriate and fulfilling decisions.

Part II

A SYSTEMATIC APPROACH FOR MANAGING YOUR COURSEWORK

Working on case studies

Introduction

Purpose of this chapter

8.1 Types of case study

8.2 What your tutors are expecting from the process

8.3 The process

8.4 A note about group work

8.5 Case study exercise

8.6 Summary

'The worst feeling in the world is when the professor 'cold calls' you to open The Case (when you're expected to analyse the issues and propose an action plan), and you have to admit – in front of your 90 section mates – that you are unprepared for this, and cannot answer.'

Nick, Harvard Business School

Introduction

Case studies are used in various guises on MBA programmes. In some programmes they are an integral part of the learning for every aspect of the course and students spend many hours drawing out the key issues embedded in them. In other programmes, they are used to supplement learning, or sometimes to examine students on their understanding of a topic.

Case studies are usually based on real examples of organizations. They vary in style and depth, but there are always important issues to be discovered within them. They are seen as excellent study aids, as they illustrate much better than any lecture can the complexities of understanding organizations.

On some MBA programmes (following the Harvard model), case studies replace the conventional lecture. Students are asked to prepare cases in advance and are selected ('cold called') in class to present their analysis. Their performance is assessed on the basis of how well they have analysed the case and how well they defend their conclusions and recommendations. Since there is no guarantee that having presented one day, you will not be asked to present again the following day, students on programmes following this model must always be well prepared and alert. One Harvard MBA graduate told us that over his two-year degree, his group of 90 analysed 800 cases! Whilst not all MBAs place such emphasis on this process, it is nevertheless an integral part of many programmes.

Purpose of this chapter

This chapter includes information on the following areas:

- Guidelines for students faced with analysing case studies for the range of purposes detailed above.
- Advice on how to go about dividing the labour available in your group and suggestions of some of the pitfalls to avoid in carrying out this process.
- Discussions of the types of case study that you might encounter.
- What your tutors might be expecting from the process.
- A step-by-step guide to conducting case study work.
- The issues of doing case studies as group work.

8.1 Types of case study

Case studies come in many formats, and can vary in length from 2 to 50 pages. Often they contain the following information:

- background and a history of an organization
- financial information
- a discussion about the leadership and culture of the organization
- an account of the strategic, operational, marketing, sales, and human resource decisions that have been made in the past.

Case studies invariably contain a number of messy or complex problems confronting the organization at the time of the case study, which you will be asked to consider in the light of the data available to you. Traditionally, these cases are written by business school researchers who have undertaken an in-depth study of the relevant organization. Often the researchers offer teaching notes (available only to tutors working with the case) suggesting ways to use the case, as well as different ways to analyse the content of these cases. Your tutors may adapt these cases for the purposes of their own sessions and use them to draw out specific learning, or occasionally they may be used for examinations.

Case studies are very often multi-dimensional and integrative. They are, therefore, likely to contain issues relating to all fields of study on the MBA programme. They may also be focused on:

- the management of change
- organizational culture
- innovation and entrepreneurship
- leadership
- information systems
- or any number of other organizational issues that are rarely seen as belonging to a single discipline.

Other cases may be written with diagnostic work with a particular subject area in mind. For example, there are financially-based case studies which enable students to focus on the organizational performance, just as there are cases which focus on the supply chain, the communication systems, the people, the technologies or the power in an organization.

8.2 What your tutors are expecting from the process

It is important to realize that analysis of a case study requires you to understand a great deal more than the initial reading will reveal.

A description of what is contained in it may be a starting point, but this is only the start of a detailed thinking process during which you will:

- extract different aspects of the text
- code and categorize your data
- choose and apply appropriate theories to this data
- engage in a sense-making process
- consider the implications of what you have found
- make recommendations or answer questions on the case.

Your tutor's expectations will vary depending on the nature and design of the case. However, one of the purposes is likely to be to give you the opportunity to apply the models and theories that you have learned during the programme to a 'real' case and, through doing this, to learn a rigorous and methodical approach to analysing organizational situations and problems. Case studies offer a safe way to learn organizational diagnostic work, as the decisions and actions that you recommend usually stay within the class. Later, it is hoped that you will be able to take these skills and apply them in the 'real world' (*see* Chapter 9).

Your tutor will therefore expect you to approach a case study as though it were a real consultancy exercise or you were a real line-manager facing this issue. The main difference is that you are unlikely to be able to speak to members of the organization (although some case studies do include role-playing exercises to enable students to learn sound consulting skills). You may be encouraged to seek additional information through library or web-based sources. However, most case studies are self-contained, with all the data you need for sound diagnosis of the situation being included in the text, and it is usually not appropriate to spend time looking up what actually *did* happen to the organization after the case was written as this may prejudice your analysis of the case. Your tutor will be looking for evidence of the way that you have engaged in this process, as well as looking at the way that you have applied relevant theory to the situation being analysed. If you can demonstrate a robust and rigorous process, combined with the application of appropriate concepts, you will be more persuasive and convincing in delivering your findings, whether this is in writing or orally.

So the process of case study work, albeit largely a paper exercise, should mirror in its approach the process of doing sound consultancy work. The features which make a good consultancy project are almost no different from the features that make good case study analysis. We would suggest that you bear this in mind as we proceed through this chapter. Whilst your tutor may be the client for your case study work, in the future you are likely to be delivering your analysis to a real client. In both cases it will be important that your methods are transparent and rigorous and that you can deliver your results with conviction.

8.3 The process

Step 1 Preparation and planning your time and approach

Case studies are often done under time pressure, which reflects the real world of management in which you will rarely feel you have enough time to accomplish the task you have been set. Time management will therefore be vital and, even before you start reading the case study, there is likely to be preparatory work that you could usefully do. Ask yourself a few questions in advance. By when do you have to complete the exercise? Do you have the option to work with colleagues on this and therefore to divide up the labour involved? If so, how will you make the best use of all team members and ensure that everyone is kept fully informed of the work being conducted by others? What deliverables have been requested by your tutors? In what format do these need to be presented? What theoretical frameworks might you need to apply in order to accomplish your task? Do you have these references readily to hand? Preparation can make a big difference to the successful completion of the task, so it is wise not to skip this step!

Step 2 Assimilating the case and breaking it down into 'bite-sized' chunks

There are various ways to do this, but the most popular is to 'skim read' the case at the beginning to gain an overall impression of what it is about. This will help you to focus on the most relevant parts as you read it for a second time in more depth. If you already know what your tutor wants you to focus on, it will be easy to highlight important aspects of the case as you read, but if you are simply asked to analyse the case without a more explicit brief, it will be harder to predict what is going to be significant for your analysis, so it is wise not to make assumptions about which aspects of the text can be disregarded.

As you undertake a second, in-depth reading of the case, it will be worth making notes as you go. One way to divide up your thinking about the organization in the light of this second reading is to look at the past, present and the future of this organization. Clearly it will be important to apply a number of conceptual frameworks and models at the next stage (Step 3) in order to address these questions in more depth, but at this stage it is nevertheless worth reflecting on the following.

- **The past**
  How successful has this organization been in the past? Why?
  What factors have influenced its successes and failures in the past?
  What problems has it faced?
  How much change or stability has it encountered and why?
- **The present**
  What is happening to this organization now?
  How successful is the organization now?
  How has its history influenced the present scenario?
- **The future**
  What problems or threats may present themselves in the future?
  How might the changing external context affect its future performance?
  How might the changing internal context affect its future performance?

As well as this first step to understanding your material, you may also notice that there are some clear themes emerging. Later you will need to read the text in different ways but for now do some quick diagnostic work to consider the following.

- **Internal environment**
  What is the organization's purpose?
  Has this changed?
  To what extent is this purpose currently supported by its infra-stucture? Are problems evident in any of the major functions?
  Are problems evident in the following areas?
    Structure
    Culture
    Systems
    People
    Technologies.

- **External environment**

  What do you know about the environment in which this organization is situated?

  How has this changed?

  How might it change in the future?

  Are there problems evident in the following areas?

  Economic

  Social

  Political

  Technological

  Governmental.

Try to *code* your notes as you read according to the topics suggested above. This will assist you with the next step of the analysis, as well as enabling you to discuss your first thoughts with others if you are engaged in group work.

Step 3 Analysis – the application of conceptual models and tools

Having asked a number of questions about the case after the second reading, you will have noticed that some of these questions suggested above are much easier to answer than others. For some you may appear to have little or no data available to you. Step 3 (Analysis) involves the application of conceptual models to develop a much more in-depth understanding of all aspects of the case. This step will allow you to understand much more deeply what is happening in the case.

You will by now have noted that there are some omissions in the data you have available as no case study can ever be complete. Do not worry about this. Even in a real-world consultancy exercise, you will always only have partial information and this information will inevitably reflect the views of those to whom you speak. In a case study exercise you are unlikely to have the opportunity to ask questions in the organization under study. This means that you will always be working with incomplete data reflecting a single view – that of the author. Conversely, some case studies 'flood you' with information, which can trap students into the classic pitfall of being overwhelmed with a lot of irrelevant data and drowning in the analysis of it, leading to frustration and lost time. Do not assume that all the data you are given will be relevant to your analysis. You must be selective.

However, many students miss the large clues available to them in the text when they start to consider some of the questions above. For example, there may not be a paragraph about the organization's culture or leadership style, but you may be able to deduce many aspects of this by considering how culture is manifested in the organization and by using models to help you to do this.

There may not be an explicit discussion about the strategy or the strategy process, but in applying some diagnostic tools you may be able to ascertain much more than you realized was in the text during your reading of it. The case may not tell you how profitable or debt-ridden this organization is and how it compares with other organizations in its industry, but by engaging in some financial calculations and ratio analysis, you will soon be in a stronger position to plot financial trends.

Step 3 is necessarily time consuming and involves the following stages:

● Identify the areas which require further analysis.
● Identify from your MBA repertoire a number of models and concepts that will help you to do this analysis (e.g. Porter's Five Forces, PEST, Value Chain, SWOT, etc.). This might mean setting up a portfolio for yourself in which you keep handy all the useful models and concepts available across the topics covered on your course. We know, for example, that often students have a cluster of analytic tools to cover questions around strategy, markets, operations, finance and accounting, human resource management, organizational behaviour, the economic environment, information systems, etc. At this stage it is best to stay broad in your analysis unless you have been asked to do otherwise, so apply as wide a range of tools to the case as you have available to you.
● Apply these models to the data contained in the case. This is better done in a group as you will benefit from the breadth of ideas, but can equally be done alone.

- Think deeply about what these models are indicating and try to form some deeper insights about the organization as a result.

At this point in your analysis you are likely to have built up a strong picture of what is happening in the organization through the application of theoretical frameworks from across a number of management disciplines. At this stage you will start to see connections between these models. The next stage suggests a systematic approach to developing these links in order to synthesize this analysis and to develop a more holistic understanding of the organization.

Step 4 Synthesis – finding connections

You may not yet have started to think about how a particular culture or leadership style is impacting on the strategy process and strategic choices made by this organization. You may have noted the financial trends but not yet connected these to the introduction of new production methods in the organization, to the arrival of a new chief executive, or to the growth of an aggressive competitor or the demise of another.

In order to draw out these patterns there are a number of possible approaches you might take.

Draw up a chronology of events

Most case studies are not written chronologically, so the connections between events may not be immediately apparent. Drawing up a chronology of internal and external trends and events should help you to spot these connections and to ask what relevance they have to your understanding of the case.

Identify trends across your analysis

Identify recurring themes that appear in more than one of your models, e.g. responses to change, impact of new technology, marketplace relationships, organizational structures, financial market conditions, etc. Ask yourself how these big themes have impacted on or been impacted by the events identified across your analysis.

Ask the Step 1 questions again

Ask yourself again the questions which you posed in Step 1, looking this time for more supporting evidence contained in your analysis. You will now have a much fuller picture with which to respond to these questions, and the questions themselves should encourage you to connect your thinking across all the analyses that you have conducted during Step 2.

Write a short summary about this organization

By now you should have sufficient data to answer the big questions about the organization, for example, where is it now? Where has it come from? Where is it going to? You will also be in a position to think about more detailed or specific questions about the case.

Step 5 Identify the key problems and issues facing this organization

Now that you have a very detailed analysis and synthesis about the organization, you are ready to move into Step 5 in which you will need to draw out and identify the key problems and challenges facing the organization.

At this stage, do resist the temptation to try to solve these or to come up with a way forward. Instead, try to draw out from your analysis and synthesis all the issues which this organization will need to resolve if it is to succeed in the future.

Try to cluster these into common areas and allow the relationships between the problems to emerge. A diagram or visual representation such as Checkland's Rich Picture technique may help you to do this (Checkland and Scholes, 1990). In rich pictures (*see* Chapter 10 on assignments), many of the key problems and the relationships between the problems are presented in visual form. Alternatively, mind maps (*see* Chapter 2) also allow the connections between higher level and sub-level issues to be identified.

Try to prioritize these problems, particularly where the solution to one is predicated on the resolution of another.

Step 6 Identifying possible future threats

Whilst many of the problems which you identify in Step 5 may be quite tangible and specific, there will also be a number of unknowns which you will need to consider, including potential threats in both the external environment and internally. Whilst these may be less controllable than the specific problems you have identified, they will still need to be taken into account in your plans for future action.

Once you have listed these, cluster them in the same way as the problems you have identified, highlighting any links between the problems and the threats.

Step 7 Identifying alternative solutions to the problems and alternative responses to the potential threats

Using the diagram or rich pictures generated to identify the problems and potential threats, start to generate two lists:

- possible solutions to the problems
- possible responses to the potential threats.

At this stage, resist evaluating your options. If you are working in a group this will feel like a brainstorming activity, the only rule being not to make judgements until all possible options have been generated.

When you have completed the generation of your ideas, start to cluster these again by theme, ranking them by importance and linking those which impact on each other.

Step 8 Deciding and recommending an action plan

In case study work, you will sometimes be asked to recommend a specific course of action. Alternatively, you may be asked to outline the options open to the organization and then recommend which of these you would select. In both cases, you will be asked to justify your proposed action plan, so all your analytic work will be important to support your conclusions. Sometimes you will be asked to differentiate between actions to be taken in the short, medium or long-term, for example, this year, next year and in five years' time.

Clearly the further away the timescales, the more you will need to take account of the unforeseen risks and threats that you have identified in Step 6.

This penultimate step will require you to imagine the future and envision the possible outcomes of your suggested solutions, as well as the risks of not achieving your plan. For example, if you decide to recommend building a new plant, this will clearly have implications for human resources, operational design, finance, etc. If you decide to downsize as a result, this will have implications for public relations, community relations, knowledge management and finance. It is clear that for each scenario you propose you need to have considered the possible implications of the decision.

Having made a full evaluation of all possible solutions and responses, you must be able to justify your choices and be clear about what your criteria for success would be. Since it will not be possible to know whether or not you are correct in your judgement, your tutor will be looking for a well-structured argument, supported by data in favour of your recommendations. Generally, the supporting arguments and data will draw on any dimensions of the organizational or business environment and should give you an excellent opportunity to demonstrate the depth and breadth of your understanding of business and management theory.

Step 9 Review

When you have completed a case study, do not miss out on the important opportunity to learn from the experience. Whether you have been working as a group or independently, do put aside time to reflect and consider how the process worked, which aspects left you feeling vulnerable, where you encountered difficulties, why these arose, how successful you were in reaching an outcome, and how the process could be improved next time.

If you felt vulnerable when dealing with a specific discipline, you will at least know that you need to spend more time reading around this subject.

8.4 A note about group work

In Chapter 3 we discuss the issues that you may encounter when working in groups. However, since case study work can be a very pressured experience, particularly if you are working against the clock, it may be worth a few words on group work specifically in relation to case studies.

You may not have any influence over the group you work with but if the opportunity does arise we suggest that you opt for a range of experience for this type of work, ideally selecting members from different functional backgrounds in order to ensure an appropriate mix of expertise.

You will discover that if you have a group to analyse a case you will need to find a balance between work done together and breaking down some of the longer tasks to be done by subgroups. This works well and can be time efficient provided that regular times are scheduled to involve the whole group and share findings. It is really important that by the end of the process each member of the group has a full understanding of the case and not just of the part that they undertook in the smaller subgroup.

Be careful not to overload any one individual and do ensure that whoever in your group has been nominated to analyse the financial data is always included and is brought up to date with discussions on other topics. We have often seen it happen that the person working on the accounting information becomes absorbed in the task to the detriment of sharing a common understanding of the rest of the case study.

If the risks are low (i.e. if it is not an examination) and the learning is more important than the outcome, try to rotate roles so that those least comfortable with a particular field or discipline agree to take on the analysis for it. This may be a little slower initially but the end results will be highly beneficial.

The boss of McDonald's is accused of running a heartless, global corporation. If only he was

Face value

Where's the Beef!

JACK GREENBERG'S favourite video clip in the wake of September 11th shows a crowd of irate Muslims shaking their fists and roaring anti-American slogans outside a McDonald's restaurant in Jakarta, the Indonesian capital. Until, that is, the aghast owner runs outside, shouting that he is a Muslim who shuts up shop each Friday so that his employees can go to the mosque. Abashed, the protesters slink away.

While the McDonald's chief executive is delighted at their response, the episode encapsulates an irony at the heart of the world's largest fast-food empire. The group is often derided as a symbol of all that is wrong with globalisation: the imposition of red-toothed capitalism and shallow American values on an unwilling world. But in fact, the firm's problem is often not that it is too global—but that it is not global enough.

Of course, with more than 29,000 restaurants in 121 countries, few companies are more international. But when it comes to management, Mr Greenberg runs McDonald's less as one worldwide company than as "an amalgamation of local businesses run by local entrepreneurs from Indonesia to France". Since taking over as chief executive in 1998, he has deliberately devolved power from the centre. By loosening the grip of his predecessor, Michael Quinlan, Mr Greenberg has rebuilt tattered relationships with franchisees, who run four-fifths of the restaurants, and has given them more freedom to decide everything from marketing to new additions to the menu.

Fine, except that this decentralisation has taken its toll on the main pillars of McDonald's brand: service, quality and cleanliness. Founded by Ray Kroc in 1955, the chain built its success less on the taste of its food than on rock-solid standards and unfailingly good service. Famously obsessive (as a retiree he would spy on his local McDonald's with a telephoto lens), Mr Kroc's saying that "if you've time to lean, you've time to clean" passed into company folklore.

Things are different under Mr Greenberg. In the United States, which still accounts for half the business, surveys show customer satisfaction falling below levels at rivals such as Wendy's and Burger King. McDonald's market share is flat, with customers defecting not only to obvious rivals, but to coffee shops and delicatessens. Add in nasty distractions such as mad-cow food scares in Europe, and the company's performance is now as soggy as some of its burgers.

On October 29th the company gave warning that profits in 2002 would miss expectations. Real sales growth has fallen (see chart), margins are shrinking and returns on investment are declining as the group pours capital into new stores. The "Mcbrand" even lost value last year, according to Interbrand, a consultancy.

Mr Greenberg is trying to stem the decline. On October 17th, on the heels of a management shake-up in May, he announced a restructuring that cuts the number of regional divisions in America from 37 to 21, and adds a new layer of management to monitor quality and impose tougher standards on franchisees. But cynics point out that the part of the business now most in trouble—the home market—is the one Mr Greenberg tried to restructure before becoming chief executive. He also bears some of the blame for slipping standards. His introduction of the "Made-For-You" food-assembly system failed to boost sales, and took staff away from serving customers. Michael Roberts, the head of McDonald's US, says that the latest restructuring will involve, among other things, simplifying a menu that started out with three items but now offers 75.

McNice Guy

One of Mr Greenberg's problems is that he is nice. For a former accountant, he is decidedly human. He likes a joke, has created a warm culture at head office in Oakbrook, Illinois, and is humble about past mistakes, admitting that "we've been distracted from running great restaurants by our growing complexity." His open manner contrasts with the single-minded arrogance that cost a former Coca-Cola boss, Doug Ivester, his job in 1999. But this amiability may make it difficult for him to face up to the tough choice now facing McDonald's: should it continue to be run as a loose federation of local franchises that can be expanded rapidly, albeit at the risk of falling standards? Or should it become a more tightly controlled global group that sacrifices some growth to maintain quality?

Although Mr Greenberg's latest move appears to inch the company down the second path, he remains gung-ho at heart, promising to open some 1,400 stores next year, the same number as this year. "We are not mature in any markets even after 46 years," he says. "New store openings will always be important and in time will accelerate."

As the network of franchisees grows, Mr Greenberg seems afraid to tinker with it. While Mr Roberts is keen on fewer, stronger franchisees with bigger territories, his chief executive has reservations about any such overhaul. "The relationship with franchisees is so important to us, it may get in the way of us being honest," he admits.

Yet the stockmarket would like to see a tighter, more centralised McDonald's. Instead of aggressive expansion, investors want McDonald's to concentrate on the profitability of existing stores. Excess cash might be better used repaying debt or buying back shares—this week, after its profit warning, McDonald's announced a second buy-back programme, this time worth $5 billion, roughly 15% of its outstanding shares. Or it might develop the group's newer pizza, sandwich and Mexican-food chains, rather than raising yet more golden arches. Yet, as Coca-Cola and others have already found, and Mr Greenberg is only now discovering, managers who excel at driving forward a fast-growing business are often unsuited to running one that is slowing down. ∎

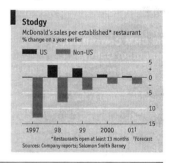

Stodgy

McDonald's sales per established* restaurant
% change on a year earlier

US   Non-US

1997   98   99   2000   01†

*Restaurants open at least 13 months †Forecast
Sources: Company reports; Salomon Smith Barney

Figure 8.1 McDonalds case – 'Where's the beef?' November 3rd 2001 The Economist

Source: The Economist, 3 November 2001 (reproduced with permission)

8.5 Case study exercise

Fig 8.1 is a very brief case study (they are usually much longer!) designed to give you some practice at analysing data in this format. The extract is taken from *The Economist* and we have chosen it because, like many case studies, it is rich in data of all kinds. We suggest that you follow the process outlined below, either as an individual or working in your group, in order to get a feel for the process you will need to follow.

- Apply the process described in this chapter to the McDonald's case described in Fig 8.1.
- Try to draw a rich picture of the current situation facing McDonald's.
- Code the topics contained in the case. Compare your list with the list in Box 1 below and endeavour to group the issues into themes.
- Anticipate possible questions and compare your ideas with those in Box 2 below.
- Decide how you might answer these questions and what MBA models would help you to analyse the questions posed and practise applying these models to the data you have in the case.

Case study exercise Box 1: Issues contained in the case (as they appear in the case but not yet grouped into themes)

Globalization

Strategy

Organizational structure (centralization *vs* decentralization)

Organizational change

Organization culture

Human resources

Leadership style

Franchising

Branding and brand value

Service

Quality

Role of founder
Competitors
Market share
Market development
Product development
Politics
Environment
Socio-economic climate
Finance
Cash management
Sales
Operations
Growth

Case study exercise Box 2: Examples of possible topics for questions

Possible strategies for globalization.

Strategic options and choices. What should McDonald's do now?

Should McDonald's become more global? How?

Fit of organization structure to organization strategy? Impact of new structure?

Changing culture of McDonald's. To what extent does its culture support its strategy?

Changing leadership styles and their impact on performance and culture.

Branding, markets and competition – successes and failures. How should it now formulate its marketing strategy?

Financial performance, cash management and options open to McDonald's. Future investments.

Operational issues facing company.

People issues facing company.

Problems of growth.

Options for growth.

By the time you have completed your analysis, you should have undertaken the following:

- Drawn out all the key issues facing McDonald's at the time that the case study was written.
- Understood what has led to these issues.
- Understood the current issues facing the company.
- Applied some key models to the data in order to structure your thinking.
- Considered what possible questions you might be asked about the case.
- Considered how you might answer each question.
- Synthesized your data so that your understanding of the marketing, strategy, human resource, financial and operational data are fully integrated in your analysis, and you understand how each is impacting on the other areas of the business.
- Considered what options are now open to McDonald's.
- Considered what you would now recommend to the company if you were asked to offer your advice.

8.6 Summary

Working on case studies can be an immensely rich experience. However, they can be daunting if you do not have a process for approaching them. This chapter has aimed to give you a process to follow. It need not be stuck to rigidly but will at least serve as a reminder of the stages that your tutors will expect you to have gone through in your analysis. By ensuring that all these steps are followed, you will find that you are much better able to defend and justify your decisions and recommendations.

Remember there are never any absolute answers to case studies and that you will be assessed on the quality of your arguments, rather than on the conclusions that you come to. Ensure that your thinking process is as transparent as possible. Good luck!

Applying theory to real-world problems – projects and field studies

9.6 Building rapport with your client and understanding the organization

9.7 Executing the project

9.8 Analysing your data by applying theory to the issue or problem

9.9 Feeding back and making recommendations to your client

9.10 Summary

'The MBA course is specifically designed so that the workload required of the individual is greater than the time available. This teaches you not only to delegate but also to accept delegation as part of a team. The value of an effective team is illustrated through your dependence on your study group colleagues' efforts and abilities for an element of your own course grades. That can be especially challenging if you do not like or respect your allocated colleagues! Swap "bonus" for grade if you are in the business world.'

Justin, Cranfield School of Management, Cranfield University

Introduction

At some point in the MBA programme, whatever its structure, most students will be expected to take their learning outside the classroom through a field study or project based in a real organization. If you are studying part-time, you may be asked to apply your learning within your own organization. Full-time students, however, are likely to be encouraged to take up opportunities offered by external organizations, which are often quite enthusiastic about offering a placement to an MBA student in the hope that they might shed some new light on a problem facing the organization.

Purpose of this chapter

The purpose of this chapter is to discuss the following topics:

- How to select an appropriate project.
- How to find and gain access to a potential organization for your project.
- How to manage your client's expectations.
- How to manage the transition from student to consultant.
- Planning your project.

- Conducting research.
- Analysing your data.
- Preparing and giving your feedback.

9.1 Choosing or finding a project

In our experience there are usually many questions facing MBA students as they approach a project period. Although the answers to your questions will undoubtedly emerge over time, you can speed up this process and feel more in control of your destiny if you become involved early on in discussing with your tutor what kind of project might suit you best. Our experience is that those students who are proactive about the arrangements for their projects, who take an active interest in where they are placed, sometimes even taking the initiative to find suitable placements, are likely to find it a more fulfilling experience. Our first piece of advice then is to *get involved early* in the decision-making process and start asking and finding responses to the following questions as soon as you can.

Some of the most common questions are as follows:

- Where will I be placed?
- How much choice will I have?
- Will I be able to do the task?
- Will I like the people?
- How useful will the project be for me in the future?
- Will it enhance my curriculum vitae?
- Will I learn something new?

In many ways the process of deciding what kind of project is most appropriate for you is similar to choosing an elective (as discussed in Chapter 7). In Chapter 7 we offer a structured approach to choosing an elective that you can adapt easily for the process of choosing a project. In summary, the three-step process outlined in Chapter 7 followed three questions. These have been adapted below for the purposes of selecting a project.

Question 1

What do I already know about the projects available and what information do I need that I don't yet have?

Question 2

What do I want from a project?

Question 3

What other factors are important to me?

Having generated a list for each question, the next step is to weight the importance of each factor, and finally to give a rating to each project based on the number of factors fulfilled by the project and the importance rating you have given to it. (For a full description of this process, please see Chapter 7.)

9.2 How to target an organization and find a project

Students sometimes prefer or are asked to make their own arrangements for projects, either because the projects on offer are limited in number, do not seem to meet the specific needs of the student, or because the students have contacts of their own that they wish to follow up. Most schools are happy for you to do this provided that the details of the project have been passed to them for their approval and that they are convinced that the learning experience on offer meets the requirements of the programme. Do, though, listen to the advice that you are offered by your tutor. They have a great deal of experience in assessing the value of projects.

We suggest that you follow a structured approach to locating and finding an appropriate project as follows.

Step 1 Decide what type of organization you are looking for

Before making approaches to potential host organizations, you will need to be clear in your mind about what type of organization and project you would most benefit from. Do you, for example, wish to gain experience in a particular sector, perhaps one that might open up future career opportunities? Do you want to be placed in a sector which is totally new to you, or to return to one in which you have previously gained experience? Do you want the experience to be in a large corporation or a small–medium-sized enterprise (SME)? How important is the nature of the project to you? Do you want it to focus on a particular functional issue or to offer a more general strategic challenge?

Step 2 Decide how to target appropriate organizations

As you become clear about what it is you are looking for, your next step is to decide how best you might target organizations that meet your profile. Do start early because this process can take some time. If you already have contacts, then of course this is the best starting point, and even if your contacts cannot help directly they may well be able to suggest other appropriate people you might target in their sector. If you do not have contacts, ask your tutor to check out whether any of the faculty in your school have conducted research in this sector and whether they are willing to put you in touch with their contacts. If all this fails, you may need to do a library or web search to uncover the names and addresses of suitable organizations to target. Try to get a contact name for each organization, preferably of the director of the division or function you wish to target. This is not as difficult as it sounds. If you telephone the switchboard of the organization and tell the operator that you wish to write to the finance director or marketing director and would like a name, they will invariably give it to you.

Step 3 Plan a structured campaign

The next step is to choose at least 12 individuals to target in your preferred organizations (you may need more later). Following the steps outlined below will not guarantee success but it will maximize the possibility of finding a rewarding placement.

1 Write a letter to the *named* individual outlining why you have chosen to target the organization, the nature of the project you are seeking, what you can offer to the project and a little information

on the MBA programme you are studying. Include something which legitimizes your letter, perhaps a brochure about your programme. You may even be permitted to write on your school's letterhead. Conclude by saying that you will phone the person in a few days. (N.B. Put a note in your diary and make sure you do this – if the P.A. answers and asks what your call is about you can genuinely say that the manager you have approached is expecting your call, and that you are phoning with regard to a letter that you sent earlier in the week.)

An e-mail would be acceptable if writing a letter is not practical, but do make sure it is personalized and not sent out as a multi-addressed mail-shot!

2 Follow up the letter with a telephone call and, if the person is not available at your first attempt, ask when they will next be in the office. If you find yourself being blocked by a gatekeeper, try phoning early or late, as you will occasionally find that directors answer their own phone at these times!

3 Once you have the attention of the person you are targeting, ask for an appointment (if this is practical). It is much easier to sell an idea face-to-face than over the telephone.

4 Make the most of the meeting – prepare an informal presentation to explain who you are, what you are looking for, how you can add value and what the advantages would be to your potential host of offering you a project.

5 Follow up after the meeting both with a letter and, where appropriate, a telephone call. Even if you do not have an immediate agreement, your follow up might just clinch the deal!

9.3 Preparing for your project and managing your client's expectations

You have found (or been allocated) a project, obtained your tutor's agreement and support and are ready to prepare for the project. You are likely to be a little uncertain still about the scope of the project, what your client expects you to deliver, what access you will have to people in the organization and who has an interest in your work beyond your immediate sponsor. These are all issues you will need to clarify early on. Ask for a scoping meeting to agree and record the following:

- the goals of the project
- timescales
- format for the report/presentation
- frequency of meetings with your client
- access to data and people
- access to workspace and facilities
- a budget or expenses for your work where appropriate.

Your school may lay down guidelines for some aspects of this list, so check first that whatever you agree is in accordance with the regulations. Where possible, ask your tutor to accompany you to the scoping meeting.

Once the scope and objectives of the project have been agreed, record your understanding of this and send it to both your tutor and your client. You may be asked to sign a contract at this stage, particularly if the client has been asked to pay a fee for your work. If there are still questions unresolved at this stage, don't hesitate to get in touch with your client – it is important that you avoid allowing misunderstandings to develop.

If your client appears to be asking too much of you in the time available (a common problem), lay out a project plan and indicate how much time you plan to spend at each stage of the process. This will enable you to indicate how far you can realistically take the project in the time available. You will need to point out any work which is going to fall outside the scope of your project and will need to be taken forward later.

Ask your client to send you some pre-reading about the organization prior to the commencement of the project. Not only will this be helpful for your understanding when you arrive there, but it also creates a good impression and indicates that you are taking the project seriously.

It is vital that you ensure regular communications with your client and your tutor throughout the project, whether it lasts one week or six months. If your timescales start to change or if you suffer a setback, contact your client and your tutor immediately. If you get into difficulties with the work itself, raise this early on with your tutor. One of their roles is to coach you through this process, but they will not be able to help you if they are unaware that you need some help.

9.4 Managing the transition from student
to consultant

Many MBA students (particularly those studying full-time) find it difficult to make the transition back to the 'real world' after so long in the classroom. Often they are slow to take up their new roles as consultants and forget that the expectations of their clients are that they will be pro-active, self-managing much of the time, and will take the initiative to resolve any emergent problems. It is advisable to try to make this transition early. For example, you may not want to give up the freedom and symbolism of wearing your student attire, but when entering your host work organization you need to consider what kind of clothing will convince them of your professionalism and competence to do the job. What you wear will, of course, be dependent on the culture and environment of the organization hosting your project, but remember that first impressions count and judgements are usually made within the first 30 seconds of an initial meeting.

Be aware that if you want to be taken seriously as a consultant you will probably need to adjust your own mindset quite significantly prior to taking on your new role. A colleague once said to a group of students about to embark on their first project, 'Are you going to *be* consultants or are you going to *play* at being consultants?' This provoked some discussion and some dissent in the room. Eventually the group agreed that they would *be* consultants and this led to a significant and important shift in their thinking that culminated in a highly successful outcome for the project.

9.5 Working in a group

A growing number of MBA work-based projects are undertaken in groups. This brings with it many additional challenges and demands even greater organization, planning, communication, feedback and sensitivity, as not only are you dealing with a client in a new environment, you are also trying to work out how to make your teamwork effective (*see* Justin's comment at the beginning of this chapter). We have already discussed group work extensively in Chapter 3 on sharing knowledge. If you have not already read this chapter and are working in a group for your project, we strongly suggest that you read it prior to embarking on the project.

9.6 Building rapport with your client and understanding the organization

We strongly recommend that you take some time to understand the organization's culture and to ask about the power in the organization. Is there more than one client whose interests you are serving with the project? Are there multiple agendas? Do you get the impression that there are political issues underlying the questions your client has posed? Has the project been undertaken before? If so, what was the result? Are any other key players involved with this project and might their agendas differ from those of your client?

It is a common pitfall to think you know what is needed and to jump straight into executing your project with little further thought about the organization or your client. It is almost always the case with any strategic project that there are multiple interests at stake and that you will only have partial information about the situation or problem, as you are likely to have been briefed from a single perspective.

As you start your preliminary enquiries, you may start to perceive that there are various agendas emerging that were not visible to you before. Playing a consulting role can be highly political, since senior managers will often wish to use your findings and recommendations in support of their own case or perspective. It is important to develop your understanding of these agendas early, and to decide where the power really is in the organization, who will determine whether your project has been a success or not, and to consider how your project might reconcile the multiple perspectives. Jeffrey Pfeffer, a writer on power and organizations once said that the definition of success at work is making your boss look good! The same often rings true for consultancy – in other words, that success is often about making your client look good. This may seem a little unethical to you, and it need not unduly influence how you go about your project, but nevertheless it is wise to take this advice into account when determining how you go about your task.

(Note: If you have already completed a module on organizational behaviour or analysis, you should be able to draw on the material in this module to start to uncover what is happening beneath the surface of the organization in which you find yourself.)

9.7 Executing the project

Data collection

Working in a live political environment, where what you do and say makes a difference, feels very different from the case study analysis that you have been practising in the classroom (*see* Chapter 8).

Another difference that you will find when you start to apply theory to 'real-world' problems is that you will not be presented with the 'facts', and indeed finding out what is happening in the organization will be one of the starting points for your project. Whereas in a case study you must trust the case study narrator and accept the story as s/he has written it, in an organizational project you will be offered many contradictory stories and will have to make sense of the situation by reconciling and interpreting what you are told. You may find there are multiple perspectives and accounts of what is happening, and even accounts of the organization's history will vary from person to person. This makes real-world interventions distinctly more challenging than doing case study analysis.

There are many ways to go about your data collection and your choice of method will, to a large extent, be determined by the nature of the question or problem that your client has posed. The process of data collection, data analysis and sense-making is, of course, iterative.

Space in this book does not permit a full exploration of the issues and you would be well advised to read a specialist publication on the topic of data collection to complement this chapter. A further discussion on research methodologies is found in Chapter 13 on how to manage and write your dissertation.

For the purposes of this chapter, however, a few of the most common methods of data collection for a consultancy project are outlined below.

The interview

The advantages of one-to-one interviews are self-evident. You can usually develop a rapport more easily than in a group, you can deal with issues of confidentiality, you can pace the discussion to suit your interviewee, and you can conduct it in their own environment in order to better understand their perspective.

Semi-structured interviews are the most popular type of interview for gathering data for organizational projects, as they will allow you to shape the interview with a number of high-level questions, but will allow you to follow up with supplementary questions according to their initial responses, without constraining you or your interviewee.

Taperecording your interviews has the advantage that you can listen to them afterwards, even transcribe them, if you have the time or resources, and it overcomes the problem of selectivity in your note-taking. However, tape recorders can be a distraction to your interviewee and may discourage openness, particularly if you are dealing with sensitive issues.

The survey

Questionnaires and other surveys are a very popular method of data collection for the obvious reason that they enable you to gather the views of large numbers relatively efficiently. Recently they have become more commonly distributed by e-mail and this has reduced the time lost previously stuffing envelopes. Surveys and questionnaires do play a very useful role in data collection and enable the generation and presentation of quantitative data. Packages for analysing and presenting survey data make the analysis of such data meaningful and, if you are conducting this form of research, you would be well advised to seek out an expert who can assist you with this.

Whilst surveys can have clear advantages for collecting and presenting data, in our view they are frequently used inappropriately for gathering information about 'soft' issues such as values, beliefs, culture and motivation. In many such circumstances, where it is *meaning* that is being sought, a quantitative presentation can be misleading.

The focus group

Originally a marketing technique designed to elicit views on products and services, this form of data collection method has become popular in management research, as it enables the views of a number of respondents to be sought simultaneously by provoking discussion within a group of participants. A focus group can be very productive, as a mix of views and beliefs can induce a more lively discussion and debate than a one-to-one interview. Difficulties arise, however, when they become dominated by a few strong voices causing others to fall silent. Only very skilful facilitation can overcome this. Another consideration is that some members of the group may be less inclined to express their views when specific others are in the room, raising important issues about the mix and level of focus group members.

Ethnographic data collection

This involves studying an organization from the inside as an anthropologist might study a tribe. Ideally, an ethnographer has a legitimate role in the organization that provides him or her with access to meetings, conversations and events which will provide live, real-time data about the situation being studied. The disadvantage of doing ethnography is that it can be time-consuming. Good ethnographies are often conducted over a minimum 12-month period! Ethical issues about observing and recording the behaviour of colleagues on the inside of organizations often also arise for those seeking to undertake ethnographic research.

Other approaches

Whilst those methods cited above are arguably the most common, many newer approaches have been developed over the last few years, particularly for collecting qualitative data. One useful method of gathering qualitative data, for example, is to elicit stories from organizational members and to extract key themes from the stories using an inductive approach (i.e. not applying any preconceived categories but looking to see what emerges from the stories). Grounded theory (*see* Strauss and Corbin, 1990) has also become a popular way of analysing data inductively by looking for emergent themes in the narratives of organizational respondents and then coding and organizing these themes. Also growing in popularity are discourse (and conversation) analysis techniques which appeal to those seeking to understand how the language used in organizations is shaped by and reflective of certain assumptions about the nature of work and society, and how this language in turn serves to validate these assumptions.

Document analysis

Students often forget the importance of documentation when analysing organizational projects. This can provide a wealth of data, both of a qualitative and quantitative nature. Try to gain access to as much relevant documentation as you can in order to compare this with the accounts given to you by your respondents. This should not just include published documents, but also files containing memos, reports, e-mails, records of conversations, letters, etc. This kind of data can often be highly sensitive, so tread with care, always obtain permission first before analysing documentation, and be careful about the sensitivity of the data you include in your report.

Secondary data collection

All of the above data collection can be categorized as primary research, i.e. the data gathering that you have conducted first hand in the organization. Secondary research is when you access the research conducted by others. This might consist, for example, of surveys already conducted in the organization that might throw light on your own study, or library research that reveals other similar studies in the field of your project which will help you to make sense of the data you have gathered. Most MBA programmes will expect students to conduct both primary and secondary research for their projects and you will normally need to show evidence of having done both.

9.8 Analysing your data by applying theory to the issue or problem

It is quite common for MBA students faced with their first 'real-world' assignment to forget all about the months of theory they have been studying and to try instead to rush straight to a solution. What differentiates you from many other possible advisers to the organization hosting your project is that you have access to a wealth of theoretical knowledge that will enable you to throw light on the problems you are facing. You must now start to think about your task in the real world in the same way as you have been thinking about your case studies in the classroom. In other words, your tutors will be looking for evidence that you can now *apply in practice* the theory you have learned in the classroom. In Chapter 8 we discuss a structured approach to analysing case studies. If you have not already read this chapter we suggest you do so prior to embarking on your project, as many of the principles for analysing your data are the same and can be applied as usefully to real-world problems as they can to a case study.

Once you have collected your data and coded it where appropriate, ask yourself what areas and issues are covered in your data and what are the important themes emerging. Having completed this phase, step out of your data in order to think about the tools, models and theories that you have at your disposal for analysing this data. Not only will identifying your models help you to make sense of your existing data, they will also help you to identify missing data that you are going to need to complete your analysis. It may be that these models lead you to return to the organization to ask further questions, talk to more people, or seek out further examples of documentation. If so, you might need another iteration round the data collection and analysis loop prior to analysing your data through your chosen theoretical frameworks.

(Note: If you are expected to turn your project into a dissertation you will not only need to feedback your findings to your client (*see* section 9.9, below), you will also need to write it up as a dissertation. Further guidance on managing and writing a dissertation, including further tips on research design, can be found in Chapter 13.)

9.9 Feeding back and making recommendations to your client

Once you have collected your data, analysed it by applying theory, models and concepts, and then made sense of what you have found, you will need to return to the objectives of the project as outlined by your client in order to check that you have met them. You will also need to turn your thoughts to what actions and further work you will recommend. Clients often tell us that at this point students become too vague, often presenting a 'wish-list' but omitting any tangible or practical recommendations for action. To satisfy your client you will need to include some practical action points. Below we have outlined with headings what your client is likely to expect you to cover in a report and/or presentation on your project. The order may vary, but do check that you haven't missed out any of these crucial questions in your report.

Contents of report/presentation to client – checklist

- Who are *you*? Where are you *from*?
- Background to the project?
- What was your brief?
- What were your timescales?
- How did you carry out your investigation?
- How did you analyse your data?

- What specific theory or ideas have you used in making sense of your data?
- What did you find out?
- What do you conclude?
- What do you recommend?
- How should your recommendations be actioned?

Making recommendations is often one of the most difficult aspects of an in-company project, so let us suggest a few golden rules for guidance.

- Always state clearly what further investigative work is needed beyond the project you have done and don't make any recommendations that you cannot substantiate from your investigation.
- If appropriate, divide your recommendations into immediate, short term and long term.
- For each problem posed or objective set by your client, state how you have addressed it, what options you have considered, why you have reached this particular set of recommendations and how each recommendation will address the problem.
- For each recommendation, say how, when and, if possible, by whom you think it should be carried out.
- For each recommendation, indicate what you think the effect of your recommendation will be, how it meets your client's needs and how your client might know if it is working.

You are likely to be required to make a presentation to your client, as well as to submit a report. Tips on how to structure your report and presentation are given below.

Tips on the report

- Your report should be written in business language.
- It should contain an executive summary.
- It should be broken up with headings and sub-headings.
- It should be clearly structured.
- Whenever possible, the prose should be broken up with diagrams or bullet points.
- Your supporting data should be attached in the appendix.

Tips on the presentation

- Try to ensure that the most appropriate audience is invited.
- Think about who will be there and what they will be hoping for.
- Aim for a style to suit their own.
- Check how much time you have been allocated for the presentation and how much time for questions.
- Try to anticipate the questions that might be asked.
- Ensure your slides are as professional as possible and take copies for your audience.

Further advice on putting together and delivering a presentation is found in Chapter 11.

9.10 Summary

In this chapter we have introduced you to the process of undertaking a project in an organization in order to apply theory to a 'real-world' problem. We have explored how to ensure that you find or select the most suitable project, how to build a rapport with your client and agree the terms of reference, as well as how to manage your client's expectations. We have then looked at data collection, analysis and sense-making as an iterative approach to managing a consultancy project. Finally, we have discussed how you might frame and present your recommendations.

How to write a good assignment

'Assignments were concentrated learning experiences in a fast-paced learning environment. As I finished each one, it felt like I had reached another milestone in terms of my understanding of management theory and this provided me with the motivation to begin the next stage of the journey towards completing my MBA.'

Mekala, University of Lancaster Management School

Introduction

It is our firm belief that doing well in assignment writing is largely about really understanding what the staff who are marking your written work are looking for in a good assignment. In the authors' years of teaching MBAs (and also in our experience as students), we have noticed that able and committed students lose marks in assignments due to a misunderstanding about what is required. This is because there is a world of difference between the kind of report written for the office (which focuses on results), and an MBA assignment (which is concerned with process). This chapter gives broad guidance about what to put in your assignment and is intended to assist in gaining good marks at an early stage. It is important, however, to note that our advice, is of necessity, general. Your own business school will provide specific regulations about your MBA assignment. *Make sure you read your own MBA regulations carefully and adhere to them.* They should take priority over the general advice given by the authors.

Before we begin the chapter, it may be useful to identify what we mean by assignment writing. MBA assignments usually take two forms. The first is a *research assignment*. This requires students to undertake research relating to a 'real-life' problem or situation. They then apply theory they have learned on their course to the problem. The second type of assignment is likely to be a *discussion paper*, in which students are required to produce an essay on a particular aspect of their programme, perhaps using existing data. This will require plenty of reading, but will not entail primary research.

Much of this chapter will also be useful to those embarking on a dissertation and we therefore suggest that this chapter is read prior to reading Chapter 13 on dissertations.

Purpose of this chapter

The purpose of this chapter is to:

- reduce the learning curve in relation to assignment writing
- provide a clear idea of what is expected.

Although section 10.2 of this chapter provides a common sense approach to all assignment writing, the examples shown relate to research assignments. A special section, 10.3, is devoted to essay-style assignments.

Whichever type of assignment you are required to produce, it is important to understand that academics are interested not only in what you have achieved, but also in how you got there, and why you have made one choice as opposed to another. The academic marker is interested in following a line of argument which is defensible – so *you must support everything you say and do with evidence and reasoning.*

This chapter is organized into three sections:

- Preparing to write your assignment
- Structuring your assignment
- Writing up.

10.1 Preparing to write your assignment

Don't leave it all until the last minute!

This piece of advice comes first, as it is one of the most important in the whole chapter. It is particularly relevant to part-time and distance-learning students, who may be given several weeks to produce pieces of written work. This might seem generous, but time races by and you can easily find yourself trying to organize your work in a last-minute panic – too late to do the required reading or order the books, too late to get hold of people you need to talk to, and too late to read through and edit your work, meaning that errors remain uncorrected. As soon as you have been given your topic, you need to get started.

Choosing a subject which is appropriate

It may be that your assignment questions are narrowly defined. However, if you have been offered a choice of subjects you will need to focus on a single topic, which might be more challenging than it sounds because it leaves the 'management' of the scope of the assignment to you. Full-time students might be given a 'broad brush' topic allowing freedom to select an aspect of their course which interests them. Part-time students, who are also practising managers, might be asked to choose research relating to a problem at work. This must be a 'legitimate' research topic – something which is relevant both to your paid work and to your MBA, as well as being genuinely worthwhile investigating. If you spend time at the start choosing an appropriate subject, you are well on your way to producing a good assignment. Suggestions for thinking of ideas for topics are given in Chapter 13 on dissertation writing and are equally relevant here. Once you have thought of a possible subject, ask yourself the following questions to help decide whether your choice is appropriate.

What is the scope of the assignment?

Select a theme that is manageable within the deadline and word limit you have been allocated. You should consider a subject which is neither too broad nor too narrow and choose a research strategy which can be implemented and analysed within your timeframe, leaving enough time to write up your results clearly and coherently. Consider whether you are looking for a subject that stretches you, or one with which you are familiar. Suppose you manage a brand of soft drinks within a company that also owns a hotel chain. Imagine you have been asked to consider the impact of recession on marketing strategies. If you can find a little extra time, you might decide to base your assignment on the hotel trade so you can learn about another aspect of your company. If, on the other hand, work pressures are heavy, it might be wiser to stick to what you know and focus on the soft drinks market. Those writing an essay-style assignment should avoid choosing a subject which is too broad. For example 'The impact of recession on global trade' would be a vast subject to cover. 'The impact of recession on the soft drinks trade in the USA' promises to be more manageable.

Does the subject of your assignment interest you?

This may sound obvious, but if you are offered a choice of topic, go for something that interests you. This is particularly relevant to part-time and distance-learning students because struggling at the end of a work day with a subject which bores you will not be conducive to producing good work. Try to avoid being persuaded by line managers (or faculty advisers) to go for a topic which seems 'suitable' but which you find tedious.

Is your choice of subject feasible?

Before settling on your subject, check that it is feasible. Does it involve any expenditure, such as travel? If so, can you or your sponsors cover this? Is your research design workable? For example, does your topic necessitate speaking to particular people and are they available? If not, could someone else substitute? Finally – especially if you are doing a work-based assignment – are you choosing something which might be politically sensitive? If so, will your line manager be open to your ideas, or is your research likely to upset people? You don't want your assignment to limit your career.

Is your subject relevant to your MBA programme?

Make sure that you don't become so involved in your reading that you go off at a tangent. The marketing lecturer may have given you freedom to choose your topic, but would expect you to retain a focus on marketing. Suppose you were doing an essay-style assignment on marketing books. If you intended to include some economics in your assignment, you would need to justify how and why this was relevant to marketing and ensure that the essay still centred on marketing books.

10.2 Structuring your assignment

Once you have chosen your assignment topic, you will find it helpful to write an assignment plan. This should outline what the chapters of the assignment will be and what you plan to include in each. Decide on a nominal word count for each section.

Below is a discussion of what the authors would hope to see in a good assignment. We will go through the assignment structure section by section, with suggestions as to what you might include in each part. To illustrate our points, we relate the discussion to two imaginary topics relating to research assignments. In order to try and accommodate a wide range of student needs, the examples will cover both the commercial and state sectors.

Examples of assignments

We will now consider the structure of the following assignments:

- **Example A**, which is an assignment on brands and marketing strategy, will focus on the impact of recession on the soft drinks industry.
- **Example B**, in which the subject is decision making and health care rationing, will relate to state spending on health care systems.
- **Example C**, on essay-style assignments, is given in section 10.3.

Our proposed assignment structure will cover the following areas:

Title page
Contents page
Acknowledgements
Abstract/Executive summary
Introduction
Literature review
Methodology
Results/findings
Analysis and interpretation of result/findings
Conclusions/recommendations
Reflections on learning (optional)
References
Appendices

Although you may not choose to present your work in exactly this format, it is important to include each of these sections in some part of your work.

Title page

Whether you are doing an essay or a research-based assignment, this should give a good first impression. It should be neatly presented and should state clearly your name, the title of your assignment, the module to which it relates and the submission date. *Everything should be correctly spelt!*

Example: title page

Could this be the real thing?

Rum and Cola: The impact of an economic downturn on the soft drinks market in North America

by
Venita Mukarji

Assignment 4
Module 543: Marketing and the global economy

Submitted Jan 17 2002

Contents page

Students often forget to include this, but it provides a useful guide to the marker. You should allow plenty of time to ensure that your contents page is *clear* and *accurate*, with page numbers correctly relating to the beginning of each section. Try and ensure that the title of each chapter is a reasonable reflection of what it contains, as shown in the example below.

Example: Contents page

Section	Content	Page
Chapter 2	*Literature review: summarizing the literature on branding in relation to market share and recession, with reference to the soft drinks industry*	14

Acknowledgements

The purpose of this section is to thank those who have made a contribution to your assignment. Acknowledgements should be brief and to the point, and should mention any member of staff or work colleague who has given assistance, such as your faculty adviser/tutor or line-manager. (This might sound obvious, but students sometimes take these individuals for granted and forget to include them!) References to family are acceptable but should be kept brief. It is not uncommon for students to mention pet animals in this section. While the authors have no particular objection to Smokey the cat getting his name in print, we would point out that this is a bit tactless if the name of your faculty adviser has been omitted!

Abstract/Executive summary

This is the reader's brief guide to what is contained in your assignment. It is not the same as the executive summary typically found at the front of a management report, because it should summarize the whole assignment not just the conclusions. The best way of tackling the Abstract is to do it once you have finished the rest of the assignment. Your abstract should be no longer than 200–300 words and needs to summarize:

- what the research problem/question was
- *why* you chose it
- *how* you went about your research
- what theories you used to inform your research and analysis – and *why*
- what you found/what your argument is
- what the implications of your findings are.

Tips for excellence!

The abstract is important because the marker will read it before tackling the main part of the assignment. It can be difficult to sum up neatly what the assignment contains but it is important to do this accurately, not only because it creates a good 'first impression,' but because it enables you to send a clear message to your reader about what to expect.

Introduction

This is your initial statement about your research issue. The introduction should set the scene and state clearly the aims and purpose of your assignment. It should be reasonably concise. The reader does not need a lengthy history of the topic, just enough information to understand *why* you have chosen it and what the background is. If you are personally involved (e.g. if you are a distance-learning student and this is a work-based assignment), it is worth explaining how you fit in to the area of concern.

Example: Introduction Assignment A

As the brand manager of a high-class soft drink, you are concerned about the impact of cheaper substitutes on your brand in the event of a recession. It would be useful for the reader to understand why this was important to you. You might outline the existing market for your drink (e.g. drunk both on its own and as a 'mixer' with alcohol).

Example: Introduction Assignment B

As a hospital manager, you want to explore whether spending on drugs in the Emergency Room (ER) can be reduced because of pressures on the budget. You decide to examine attitudes to spending on drugs in ER because this will help you to understand how decisions on expenditure are made now and may help you develop a model for rationing spending in the future.

Literature review

In this section, you should be reviewing the literature, theories and academic frameworks that are *relevant to your assignment question*. It is important to explain *why* you have chosen certain theories and *how* they relate to what you are doing. At the same time, you might need to justify *why* you do not intend to consider certain frameworks. Examples are given below to illustrate this.

Example: Literature review Assignment A

As the brand manager of an expensive soft drink (Jazz), you might begin by considering the general literature on branding and market share. You might want to consider the general literature on building successful brands and could argue the importance of considering a range of factors. You might then turn to the general literature on the impact of substitutes on market and might discuss a useful model such as Porter's Five Forces model (De Wit and Meyer, 1998: 347).

You could then turn to literature relating to the impact of economic downturn on brands and relate this to your own brand. In this instance, we will imagine you have uncovered evidence which indicates that high quality spirits may retain market share during recession. At this point you may introduce the concept of a strategic alliance with a relevant alcoholic drinks manufacturer. You might therefore discuss the literature on strategic alliances, introducing a framework such as the one used by Brassington and Pettitt and reproduced here in Figure 10.1.

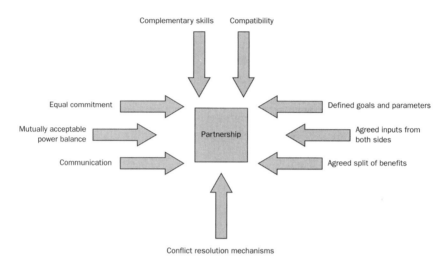

Figure 10.1 Framework of strategic alliances

(Reproduced with kind permission of Brassington and Pettitt, 2000)

Example: Literature review Assignment B

Your assignment relates to decisions about drugs expenditure in the Emergency Room at your hospital. You wish to balance the needs of patients in a situation where new and more expensive (but effective) drugs are constantly entering the market, against a limited budget. In effect, your job is to ration health care. You wish to do this in an ethical manner.

Your literature review might be split into three sections.

The first of these might consider broad literature on decision making and priority setting within the public professional services (for example, Bryson, 1993), focusing on specific models and frameworks to assist with decision making.

The second section could consider the literature on health care rationing. What frameworks for rationing heath care already exist? You might discover various attempts at developing mathematical formulae, league tables or other technical systems for making decisions.

In your third section, you might evaluate mathematical formulae for rationing spending on health care in the light of the literature on health ethics. Some writers on health care (e.g. Malik, 1994) believe that while hospital managers would like to find mathematical 'solutions' to the problem of rationing health care expenditure, those that exist are unhelpful to staff faced with the real choice of denying expensive but maybe life-saving treatment to a patient.

Tips for excellence!

A distinction answer does not merely describe the literature that has been read but *evaluates* this in relation to the subject matter. You might, for example, argue that Malik's work is useful because although it does not provide an 'answer' to the problem of agreeing health care priorities, it acknowledges and discusses the related ethical dimensions. Having recognized its helpfulness, Malik's work should then inform what happens in your assignment and should be related to the analysis of your results.

Methodology

This section is important for a research-based assignment. It enables you to explain what approach you took for your research and what methods you used. Students often fall down in this section of their assignment because they are unsure what the marker is looking for. It is worth noting that some MBA programmes do not provide formal research training at the assignment stage because teaching on 'how to do research' is reserved until the dissertation stage. If this is so in your case, we suggest doing some reading about the various techniques on offer. Some helpful texts on research design are suggested in the section on additional reading at the end of this book (*see also* Chapters 9 and 13).

The methodology section is your opportunity to explain *how* you went about your research and *why* you chose the methods you did. Is your research qualitative or quantitative? *Why* did you select this approach? Although your reader will not expect you to be deeply philosophical at this stage in your MBA, she or he will seek some *justification* for your choice of methodology. Remember that academics are interested in process, not just the results. You will get credit for the amount of work you have put into your data collection – *but only if you explain to your reader exactly what was involved.*

Example: Methodology Assignment A

As the manager working in the soft drinks industry, you have already discussed the potential for expensive spirits to maintain sales during periods of recession. This has inspired you to consider forming a strategic alliance with a spirits manufacturer in order to promote your soft drinks as 'mixers'. Arguably, you do not need to undertake further quantitative work because you already know the strength of the hard liquor market. You decide to test out colleagues' reactions to the idea of a strategic alliance by undertaking some qualitative research. In this instance you decide to facilitate a 'brainstorming' session. You would need to outline *why* brainstorming was an appropriate choice. It would be advisable to show that you have undertaken reading to help you manage the session effectively (e.g. Hill, 1991). You would need to explain who you included in the session and *why*. The assignment marker would hope to learn *how* you managed the session – for example, did you have a formal 'incubation' break? Did you contribute, or stick to a facilitatory role and *why*?

Example: Methodology Assignment B

You decide to survey two groups of medical policy makers within your hospital to identify their decision-making methods. You focus on doctors and accountants because the first group spend the money which the second group are responsible for managing. You hope to discover whether attitudes vary between the groups.

You decide to send out a written questionnaire in order to allow respondents to remain anonymous. You accompany this with a letter explaining who you are and what your research is for, which you include in the appendix of your assignment. You undertake detailed reading on questionnaire design, which you summarize in the Methodology section. You explain who you chose to survey and why, as well as outlining the reasoning behind your chosen survey method.

Real-life example: Methodology

Example 10.1

Ismail Weinstock worked for a pharmaceutical company and produced an assignment on the allocation of R&D funding. He interviewed six board-level managers, each for two hours. He spoke to them because they were the key decision makers in terms of R&D spending. He went to great lengths to track them down and travelled across continents to do so. One manager was interviewed on a Sunday at his Swiss skiing lodge. Ismail characterized this in his methodology section in one short sentence: '*I interviewed six managers.*'

Fortunately, his faculty adviser saw a draft of Ismail's work before it was handed in. She knew how hard he had worked and advised him to explain who the interviewees were, why he had chosen them and how he had gone about the interviews, detailing where and when these had taken place. Ismail was given credit for the work he had put into his research and was given a good grade – which would not have been earned had the marker read the original version of his methodology section.

Results

If you have invested time and effort in your research, the results section will be challenging because you will be unable to include all the data you have collected. You will have to make hard decisions about what is relevant and what is not. One simple but helpful way of doing this is to return to your original research question. Then ask yourself: '*Does my data help me answer this?*' You can then organize your data into three (real or imaginary) piles: **1** yes; **2** no; **3** not quite, but it moves the question forward and so should be included.

As you report your findings, ensure that each section is logically sequenced and that you precede it with a helpful heading. If your research is quantitative and you wish to include figures and tables, make sure that you explain in the text *why* they are there and *what* they are supposed to show. Don't just assume that your reader will work it out!

Do not be afraid of including negative as well as positive data in your results section – you do not need to 'sell' your solution to your marker in the same way as you might have to convince the company Chair that your work was flawless. Academics are interested in *process* and therefore would wish to see a truthful account of your *actual* results, rather than an embroidered version of what you hoped they would be. Try and make the results section interesting and convincing by backing up your reported findings with evidence. This may sound obvious, but it is all too easy to make a sweeping statement such as 'Emergency room (ER) has a problem with overspending on drugs' without providing *specific evidence* for your reasoning. You should give figures to support what you say if your work is quantitative, or quotes if your work is qualitative.

Example: Results Assignment A

Imagine your results were inconclusive. Some of your colleagues were keen to implement the plan to form a strategic alliance right away. They asked you to develop your proposal. Others were more hesitant. The Head of Economic Intelligence thought it unwise to rely on existing research and preferred to commission some of her own. She worried about the risks of strategic alliances. Don't try to 'cover up'. Your marker will recognize that not everything goes to plan during 'live' research and will be interested to know how you dealt with the problems that arose.

Example: Results Assignment B

Quantitatively, you might report that the problem of overspending on drugs in ER cannot be ignored because 85% of the budget is spent in the first nine months of the year, leaving a shortage of funds for the remaining quarter. You would include a table or graph to illustrate this point.

The results of your survey indicate that attitudes vary between groups. You learn that management accountants are more likely to use mathematical formulae than doctors (who make decisions on an intuitive 'case by case' basis). Suppose you had included a 'comments' section at the bottom of your survey. If you wished to demonstrate the conflicting views of doctors and accountants you might include anonymous quotes to illustrate your point:

'The doctors just don't understand – the budget is finite and if it all goes in the first six months, then people who get sick towards the end of the financial year will get a raw deal.'

(Management Accountant)

'Finance people have no idea how it feels to be faced with a very sick person and not give the best treatment available. Of course we are not thinking about money in this situation – just about what we can do to give the best outcome for the patient.'

(Doctor)

Analysis/interpretation of results

In this section you are going to draw together the findings from your research and the relevant theory you have discussed in the literature review. How does the literature relate to what you found? Does it bear out, or conflict with, your results? You should acknowledge the problems and constraints, as well as the successes. It is important to back up your own opinions with evidence – whether from the literature or your own research.

Example: Analysis Assignment A

You identify a brand of rum (Wild Monkey) which mixes well with Jazz, your soft drink. Using the Brassington and Pettitt framework to help your focus (illustrated in Figure 10.2), and writing your text around this, you undertake analysis which suggests that a strategic alliance with Wild Monkey could be beneficial for both products, but carries risks and raises some questions. You argue that an alliance would be beneficial in marketing terms whether or not the economic climate is good. However, you can outline options for further research on the hard liquor market in times of recession.

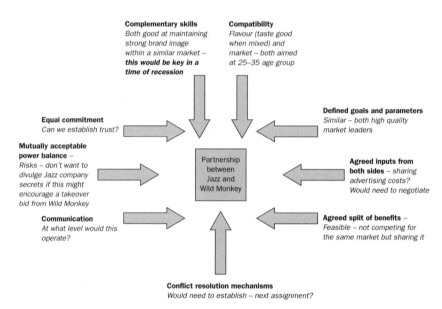

Complementary skills
Both good at maintaining strong brand image within a similar market –
this would be key in a time of recession

Compatibility
Flavour (taste good when mixed) and market – both aimed at 25–35 age group

Equal commitment
Can we establish trust?

Mutually acceptable power balance –
Risks – don't want to divulge Jazz company secrets if this might encourage a takeover bid from Wild Monkey

Communication
At what level would this operate?

Partnership between Jazz and Wild Monkey

Defined goals and parameters
Similar – both high quality market leaders

Agreed inputs from both sides *– sharing advertising costs? Would need to negotiate*

Agreed split of benefits –
Feasible – not competing for the same market but sharing it

Conflict resolution mechanisms
Would need to establish – next assignment?

Figure 10.2 Adapted framework of strategic alliances
(Adapted with kind permission of Brassington and Pettitt, 2000)

Example: Analysis Assignment B

In the analysis section, you would *analyse* the results reported above and *relate them* to the literature you had reviewed. In discussing your findings, you could consider the lack of consistency between the doctors and accountants and interpret this as demonstrating a lack of understanding. You could defend your reasoning by including some quotes (as suggested above). You could relate your results to the literature previously discussed and show how this sheds light on the situation. You might argue that the accountants favour mathematical formulae as a decision-making method because they are responsible for the health care budget but are not faced with making decisions about individuals. The doctors, on the other hand, deal directly with the needs of particular patients. You might reason that their dilemma concurs with the observations of Malik (1994) which portray the rationing of health care expenditure as a complex problem which cannot be dealt with by applying a set of principles. However, you might then criticize Malik because it focuses on the problems inherent in health care rationing *without putting forward a solution* for dealing with this. You could then refer to the work on public professional service and decision making to see if a practical way forward is offered. However, you might also argue that, while your research has helped you understand the situation better, it cannot provide a 'solution'. You could end the chapter by acknowledging the limitations of the research and identifying how the issues could be taken forward in future.

Tips for excellence!

An excellent assignment develops a *logical argument* in the Analysis section, drawing on evidence from the literature and research findings. Personal opinion may be offered but in every case must be backed up by *evidence* from either research findings or the literature to justify the line of reasoning. At each stage, the writer would anticipate the question 'why?' before it enters the mind of the marker and offer a defensible explanation.

Conclusions/recommendations

Look up the word 'conclusion' in your thesaurus and you will find its meaning given as 'the end', 'the finish', 'wrapping up'. This ought to provide a clue as to what goes in this section, which should be 'short and sweet'. It should summarize and tie up the arguments discussed in the analysis section of your assignment. If your topic is related to a 'real' work situation, you might include recommendations. You might also make suggestions for further research in the topic area. There should be no new concepts introduced in your conclusion section!

Example: Conclusions Assignment A

Following research into the soft drinks market, the formation of a strategic alliance with a high quality spirits manufacturer was considered as a strategy for maintaining market share of a soft drinks brand during recession. It was agreed that further research into the alcoholic drinks industry was needed before this could be taken forward. However, it is also worth considering this as an option for the future whatever the economic climate as it could be beneficial for the marketing of both products.

Example: Conclusions Assignment B

The research has provided an insight into the different perspectives of the two policy-making groups. This has enhanced understanding about why solutions to health care rationing are difficult to find. One recommendation might be for the results of the research to be fed back to both groups, which might improve understanding between them and may help in the search for a solution.

Reflections on learning

Some MBA programmes might ask you to reflect on your learning at the end of each assignment; others may expect your reflections on the research process to be embedded in the text as you go along. If you are writing a section to reflect on your learning, this will be expected to be short but insightful. It will provide you with a formal space to reflect on the experience of researching and writing your assignment, the usefulness of the theory you have applied and the validity of your findings. You might also consider what you have learned about the process and how this might influence your approach another time.

References

In an academic piece of work, it is important to acknowledge all sources drawn upon or cited in your assignment (you will receive credit for doing this). If you wish to refer to texts which you do not mention in your assignment but which were useful, you would include these under a separate section entitled Bibliography. There are various conventions for the way in which references should be presented, so check your MBA regulations to see which style you should use. In particular, check on how you are supposed to present references which you have printed off the net. The references in this book are cited in what is known as 'Harvard' style, which is probably the most commonly used method. When citing a reference after a quotation, you would simply write Dicken (1998: 40), placing the author's name, year of publication and page number in brackets. If, however, you want to paraphrase Dicken, then you would write, for example:

'Dicken (1998) has suggested that there has been a marked shift in Japanese foreign direct investment in the manufacturing industry...'

placing the year only in brackets. Your reader would turn to the reference section at the end of your assignment to find Dicken among other authors, neatly listed in alphabetical order. They would then have no trouble in tracing the relevant book. Suppose you acknowledge a journal article and a book by Dicken, they should look, respectively, something like this:

Example: References

Dicken, P. (1988) 'The changing geography of Japanese foreign direct investment in manufacturing industry: A global perspective,' *Environment and Planning*, 20: 633–53.

Dicken, P. (1998) *Global Shift: Transforming the world economy*, Paul Chapman, London.

Appendices

Use this section sensibly and sparingly – if at all! The Appendix is not the place for important findings, nor a means of extending the word limit, nor a way of putting off the decision about whether you should include or leave out something. The type of material included here should be limited to copies of questionnaires, letters written to interviewees, etc., which do not need to go in the text, but which might be useful to the reader. Appendices must be referred to in your text and should be carefully numbered in the order to which they are referred. As a guide, the text should make sense to your reader without the appendices, so these should be viewed as providing supplementary data only.

10.3 Essay-style assignments

Most of the advice offered above is relevant whether you are writing a research-based or an essay-style assignment. If you are writing an essay, it is still just as important to ensure that articles quoted in the literature review relate to your main topic of discussion. References should still be clearly and neatly cited and your work should still have a beginning, a middle and an ending. You will therefore still need an Introduction, outlining your aims, and explaining why your topic is relevant.

For example, if you were a full-time student, whose previous job was as a bookstore manager in the USA, you might wish to consider the impact of internet companies on the book trade. You will still need to think carefully about the scope of your assignment. Will your focus be the United States, or will it be international? Will you be talking about all books, or just one type, for example, children's literature? And so on. You will need to justify your choices. As above, your Literature Review should outline the relevant theories and journal articles on your subject matter. You might consider the impact of pricing and the economic climate in relation to competitive advantage. You may still need to include a section in which you discuss your approach, but the approach will be slightly different since you have not been doing empirical research. Perhaps, for example, you are going to focus on the economics of market structure, pricing and competition. You still need to explain how you have used the various frameworks you have selected to inform your work. If you are going to produce a figure about demand, pricing and market structure, you will need to explain where you found this, how it is relevant to your work and to what you intend to apply it. Your results and analysis sections should focus on what you have found and how the relevant literature relates to this, as above.

How the essay-style assignment differs from the research-based assignment

Where the style of the essay-type assignment differs most from the research-based assignment, however, is probably in the way your discussion is presented. The marker will be looking for the thread of an *argument* which can be developed through the piece of work.

Suppose your chosen essay title was as follows:

Title: To be, or not to be?
The future for traditional bookstores in the light of new internet companies

Your marker would want you to develop an argument for each of two or three scenarios. These could include the demise of bookstores altogether, the demise of internet companies selling books online, or the possibility that both will exist side by side for some time. You would need to explain *why* you chose these scenarios and to follow up the argument for each example right through your assignment. This means not getting so carried away with one example that you forget the others. It would also mean approaching the matter with an open mind. While it might be tempting to exclude the possibility that your own trade will be subsumed by the competition, an MBA assignment provides an opportunity to be more realistic and to look seriously at threats and competition.

It is important therefore to give a 'fair hearing' to each scenario offered as you apply the frameworks you have chosen. In your analysis section, you have an opportunity to evaluate the likelihood of each scenario in the light of your application of academic theories and evidence about the relevant market. You are then at liberty to draw some conclusions of your own – but be clear about where the opinions stated are your own and make sure that you defend them by using examples from your own analysis. As advised above, your conclusions should draw together the arguments that have been developed as your work has progressed. They might include some recommendations about how bookstores can retain competitive advantage in the light of competition from internet companies, but the Conclusion is not a forum for introducing new information.

10.4 General guidelines for writing up your assignments

As well as explaining what to include in each section of the assignment, we thought it worth offering some general tips on writing up which you might find helpful to refer to as you get started.

Reference your sources – and never plagiarize

Plagiarism is a serious offence in all universities. Before discussing serious and intentional plagiarism, it is worth noting, again, that all quotations and ideas *must* be fully referenced in your text. Occasionally, students are tempted to try and pass off somebody else's work as their own. When things are tough, 'borrowing' someone else's work to get you through might seem like a good idea at the time. This generally means either copying from a friend's assignment, or getting hold of 'ready-made' assignments from the net. There are even companies who will offer to provide you with a supposedly 'tailor-made' piece of work. Beware. The price paid for this could be higher than you expected. If someone is prepared to help you 'cheat' the system, it probably means they will be prepared to cheat you, too! Usually, students who plagiarize others' work do this for four reasons:

- they are behind with their own work
- it seems like an easy option
- they do not expect to be found out
- they have no idea how seriously this will be regarded by their institution.

Faculty staff usually sense when students are cheating and those who plagiarize are likely to be caught out. We have both dealt with students who have copied work from other sources, without referencing the sources. In every case, it has been evident that those concerned had no idea how seriously the 'crime' of plagiarism is regarded by universities. We therefore cannot stress the following point strongly enough.

Universities consider intentional plagiarism in the same light as a firm of accountants might look at fraud. Students who are guilty of intentional plagiarism will find themselves in serious trouble with their faculty and may be asked to leave their MBA programme. Even if you are desperate, it is not worth the risk of trying to pass off work as your own when this is not the case. Don't be tempted.

Two real-life examples: Plagiarism

Example 10.2

Lecturer Joe Preston was disbelieving when an MBA student presented an erudite assignment relating the works of Karl Marx to the hotel industry. Out of curiosity, Joe undertook a search on the net. He soon discovered that the basic components of the assignment were posted on a 'help with your MBA' site. His student had copied what was there and adapted it to the hotel industry.

Example 10.3

Anna-Mei Lester, a regular marker on MBA programmes, experienced a sense of *déjà vu* when reading through a strategy assignment. She asked the MBA secretaries to look through past assignments. It was discovered that the assignment in question was directly copied from one submitted four years earlier by another student (which explained why it seemed so familiar!).

Joe's student was given a written warning and asked to re-submit the assignment in his own words, but remained on the MBA programme. However, the student who had copied her assignment from a colleague was asked to leave her MBA programme and excluded from the university.

Consider the language you are using as you write

Jokes are rarely a good idea because not everyone shares the same sense of humour as you do. In an assignment about car manufacturing, you can imagine what the authors' (of this book) views might be in relation to jokes about women drivers! *Remember that your work might be marked by someone you do not know*, so avoid anything that could be construed as racist, sexist or otherwise insulting to your marker (even if it is supposed to be funny). If thoughtless comments are offensive to your reader, they are unlikely to create a good impression. This advice might sound obvious, but below are some examples that the authors have (yes, really!) come across in assignments over the past 12 months, which demonstrate how easily students can offend by failing to think through what they are writing down.

'The book was quite helpful but unfortunately it was full of Americanisms.'

'Our nice boss was replaced by a typically brash Australian.'

(in a strategy essay) *'Irish Joke:*
> **One Irishman to another**: *How do I get to Limerick?*
> **Second Irishman**: *Well, if I was going there, I wouldn't start from here...'*

'The writer of the article was an anally-fixated failure frightened at birth by a passing statistician.'

Read your work through thoroughly before handing it in

As MBA students, both authors had classmates who stayed up all night to complete their assignments.* Although you might scrape by with this technique, it is not the way to achieve good marks. Leave yourself plenty of time at the end of your assignment to read carefully through what you have written. Have you cited references correctly and in every case? Does everything make sense? Have all the pages been printed off neatly and are they all in the right order? Is your spelling accurate? Don't rely too heavily on a spell-checker, which might not pick up that you have reconsidered the role of the marketing *manger*, as opposed to that of the marketing *manager*. Don't leave it to your secretary to do this – do it yourself.

Tips for excellence!

*Neither Caroline nor Sharon ever did this! As students, no matter how tight the deadline, we always allowed plenty of time at the end for reading through and editing. If possible, it is useful to put your completed assignment to one side for 48 hours before the final read through – it is much easier to spot mistakes if you have returned to your work after taking a break from it.

10.5 Summary

This chapter has provided some basic suggestions on how to write a good assignment. Although the chapter 'stands alone', to get the most benefit you should read it in conjunction with the section on study skills. We summarize the chapter with five final tips – key points to remember when you begin your next assignment.

- Think carefully about your choice of topic. Is your subject 'legitimate'? Is the scope of it about right?
- Back up statements with evidence. You must be able to defend what you say.
- If your assignment is research based and things go wrong, you can still write a good assignment by analysing how you could have done things differently. Academics are interested in process as much as results. This is your opportunity to question things and push the boundaries. Have a go.
- Never, ever, try to pass off someone else's work as your own.
- Don't worry. It gets easier as you go along.

Making a successful presentation

'The main thing about presentations is to be well prepared, then try and forget about yourself and concentrate on communicating with your audience. If you stand up straight and face your panel of examiners (not the projection screen), and don't fidget, that will make a big difference to giving an impression of confidence.'

Janet, University of Hull

Introduction

At some point during your MBA, you will face the prospect of giving a presentation. Full-time MBAs, in particular, will be required to present their work many times to both their peers and faculty staff. However experienced they are at doing presentations in the workplace, many MBA students find presenting their ideas to an academic panel daunting and are often disappointed by the marks received. This is unsurprising because when speaking at work, most managers are on familiar ground and know what is expected of them. They are probably talking to people they know and will be practised at anticipating the likely reaction to what they are saying.

Presenting to an academic panel is different. For one thing, unless you are applying for a job, it would be unusual to be given a 'mark' for your performance at work. Feedback (if any) would probably be given discreetly on a one-to-one basis. However, at an MBA presentation, the academic panel might ask challenging questions and probably won't hesitate to give blunt (and often public) criticism of the content and style of your performance.

Purpose of this chapter

This chapter focuses on how to improve your presentation skills and enhance your grades. This general information is relevant to everyone, but at the end of the chapter, special attention will be given to two groups of students who find MBA presentations especially

difficult. The first group are those who are not nervous about presenting because they are confident and accomplished speakers, but
who are surprised and disappointed to find their marks lower than
expected. The second group are students who suffer from extreme
nervousness and who do badly because fear gets the better of them
on the day. We will attempt to analyse why these groups do less well
than they had hoped and provide some hints to help these students
improve their presentation grades.

This chapter will give an overview of what is expected in an MBA
presentation and will explain the following.

- What criteria the academic panel will judge you by.
- How to meet their expectations.
- How to make the most of your performance.
- What to include – and what to leave out.
- How to answer questions.
- How to improve marks if you are a confident presenter who fails
  to do as well as expected.
- How to overcome nerves.

11.1 Putting your presentation together

It is more than likely that your presentation will be related to a piece
of work you have already done – perhaps an assignment or, at the
end of your programme, your dissertation. Your job is therefore to
show your work in the best possible light by including the same
sorts of information that would comprise an assignment (*see*
Chapter 10), explaining clearly the following:

- what you did (giving background and context)
- why you did it
- how you did it
- how it relates to the academic theories you have learned on your
  MBA course
- what your conclusions are.

The first point of note is that the conclusions are only a very small part of your overall presentation. This is because, as noted in Chapter 10 on assignment writing, academics are interested in *process* (*why* and *how* you did things) as much as the end results. Thus, a description of your methodology and the academic frameworks you have used should form the core part of your presentation, since this is where the main body of marks will be awarded.

11.2 The content of your presentation

What to include

The purpose of your presentation is to tell a story. Like your assignments, it should have a beginning, a middle and an end. In order to manage the expectations of your audience, it is useful to begin by introducing yourself and giving a brief summary of what you are going to say (rather like the Abstract section of an assignment). An example of this is given below.

Example: Introduction to an MBA presentation

'My name is Ali Smith and I am a marketing manager for a major pharmaceutical company. The purpose of my project is to evaluate, on behalf of the marketing director, three new products (A,B and C) which are in the research and development phase of their lifecycle. It is proposed that the company should invest heavily in one of these new drugs with a view to bringing it to market in two years' time. They have sought my opinion on which product to choose.

'In view of my role in the company, I am considering this from a marketing angle. I have therefore applied Michael Porter's "Five Forces" model and "value chain" analysis to each product to help gain an understanding of the implications and opportunities of bringing the products to market. This has enabled me to rate the products in order of preference, product B being the preferred option. I am aware that there are other issues to be considered in the choice of which product to launch and two of my colleagues are looking at the medical and economic angles of the products, with all our reports due to be concluded and considered by our respective directors next month. However, my remit is marketing and that will be the focus of my presentation today. I will therefore explain to you my research methods, why I chose Porter's work and how this helped enhance understanding of the situation…'

In the above introduction, Ali has clearly outlined his subject and has managed the expectations of his examining panel, who now understand who he is, what he is doing and why he is doing it.

Having clearly and briefly explained your subject matter, you can then go on to outline your research problem in more detail and discuss your approach. Your presentation should comprise a summarized version of the kind of information contained in an assignment (*see* Chapter 10). Your academic panel will probably expect you to include the following information.

- What you did for your primary research (e.g. interviews, surveys, etc.) and your *reasons* for your choice of methodology.
- Which academic frameworks you used to analyse your problem and why you chose them.
- How you applied the academic frameworks to your research.
- What you learned from this.
- What your conclusions were and, if relevant, what your recommendations are.

Tips for excellence!

At the end of your presentation, if you feel you can do this comfortably within the timescale, you can gain extra marks by summarizing again what your project was about and then reflecting on your learning process. Your panel will be interested to hear how your project helped you develop your own skills and what you have learned from it. What went well and what went less well? Is there anything you would do differently another time and, if so, why? Remember that your academic panel will be interested in the *process* of what you did and, if you can demonstrate that you have reflected on your work and learned from it, this can gain you extra marks.

What to leave out

Background to your work

Students are often tempted to spend too long explaining the background to their work. While this is important because it enables the examiners to situate the presentation in context, it is unnecessary to spend more than a few minutes (at most) on this, as it wastes precious time that could be better spent exploring the details of your academic approach. You will note that in the example given above, Ali Smith has included brief but sufficient details. If the panel wish to know more about the background to his work, they can always put questions to him at the end of his presentation.

Jokes and wisecracks

As with assignments, we would advise being careful about making 'wisecracks' unless you are certain that they are funny, relevant to the subject of your presentation and unlikely to offend anyone. It is possibly even more important, in a spoken presentation, to be careful what you say, and about whom, because your examiners are sitting right in front of you and might challenge you about it in the end. Never, ever make jokes which refer to others' ethnicity, sexual orientations, gender or politics in a presentation. Universities might appear on the surface to be traditional institutions (and in some respects, they are) but they tend to attract 'non-conformist' staff. Thus, quips about political correctness, feminism, Marxism or anarchism might fall on stony ground because even if your smartly-dressed academic panel look like a bunch of well-heeled capitalists, appearances can be deceptive. You might be sitting opposite the author of the seminal text on one of those very subjects, and jokes about it will be unlikely to create a good impression. One good 'real-life' example of this would be a student working on a marketing campaign who referred, in disparaging tones throughout his presentation, to environmentalists (of whom there are many in universities) as 'bunny huggers'. The panel was not impressed.

If your subject matter means that you need to refer to issues that could be 'sensitive' (single-sex relationships or children with learning disabilities, for example), try to ensure that you are using language which is up-to-date and acceptable to the group under consideration.

11.3 Presentation style

The practice of whether marks are awarded for presentation style varies between schools. It is worth being sure what the requirements of your school are before you begin so that you can prepare well in advance for this. Even if you are not assessed on this, however, the clarity of your message will undoubtedly be affected by the way that you conduct your presentation. Although this may sound obvious, one of the main pieces of advice we can make with regard to improving presentation skills is to ensure that your visual aids are accurate, readable and not overwhelming in number. Students are often tempted to cram in far too much information on each slide, meaning that text is small and difficult to read. With the advent of software developed especially for presentations, it is also tempting to use exciting devices to decorate slides including lights that flash down the side of the text and swirling backgrounds that change colour. Do not do this just for effect. Any enhancements to your slides should be designed to draw attention to your main points, not to show you are a master at Powerpoint. Whatever layout you choose for your slides, remember that your audience must be able to read them and that too much background activity is a distraction.

Although your computer has a spellcheck, it is important to read through your slides carefully and in good time to ensure that you have not made careless mistakes, which will be embarrassing when they are projected large on the screen. One of the authors sat through a presentation about hospital services where the word 'health' was spelt 'heath' throughout. (Your spellcheck will not spot this one!) Make sure that your slides appear on the screen in the right order and that there are not too many of them. As noted in 11.4 below, you will be working to a time limit and it will be difficult to keep within this if you have too many slides. One of our students tried to get through 43 slides in a 30-minute presentation, leaving both herself and her audience breathless, not to mention getting no further than her introduction before running out of time.

As a general rule, use few slides and allow yourself time to talk about them, so that your panel have time to look at you and not just a screen. With fewer slides you will have time to make eye contact and engage their interest.

11.4 Practice makes perfect

Before they are seen on Broadway or in London's West End, new plays are previewed in provincial theatres. This gives both the cast and the production team a chance to review and amend the show before they perform in front of the critics. It also provides them with the opportunity to be precise with their timing. Anyone who has taken part in amateur dramatics will know that shows which are under-rehearsed tend to overrun because the actors are slow to react and pick up their cues.

One of the main reasons why MBA students do badly in presentations is because they have not rehearsed what they are going to say. Their presentations, therefore, take too long, or important points are forgotten. Most management or business schools will treat formal presentations in the same way as they treat exams and, when your allotted time is up, you will be stopped whether or not you have finished speaking. There is, therefore, no substitute for running through your presentation several times *out loud* to make sure that everything is going smoothly and you can finish on time. If possible, it is helpful to practise in front of someone else who has an understanding of what is required for an MBA. Other students or, if you are on a distance-learning programme, a colleague, could help critique your presentation, both in terms of whether the content is comprehensible and whether the style of the presentation is appropriate.

11.5 Communicating with your audience and dealing with nerves on the day

Practising before the event takes place is even more important for those who are nervous than for those who are confident about their presenting skills (though it is advisable for everyone). For those who find presentations difficult, practice will help improve your audience communication skills. If possible try and imagine that your academic panel are friendly people who you know well and that you are telling them a story. That means raising your head and looking at them when you speak, rather than being tempted to lower your head and read from a script. *Try and concentrate on what you are telling them, rather than worrying about what they might be thinking.* There is no need to use complicated management jargon (especially if this relates to your particular line of business and won't be generally understood). Just speak plainly and clearly and try not to rush your words. People who are shy about making presentations tend to drop their voice and their eyes at the end of a sentence, meaning that they appear to be mumbling. If this is you, be aware of it and try to 'practise' yourself out of the habit. Don't be tempted to make the classic mistake of looking at the overhead screen instead of your audience, so they see only the back of your head and cannot hear you properly. It will also help if you consciously straighten your shoulders, stand tall and breathe deeply and evenly.

If you can present an image of quiet confidence during your pre sentation, this will help to convince your audience that you are competent and well prepared. You will notice that news 'anchors' sit straight, look up at the camera and keep their hands still. Nervous MBA presenters, on the other hand, often give the game away by unconsciously playing with keys, ties, jewellery and hair, which can be distracting for the examining panel. Try to minimize your tendency to do this by taking the following advice:

- Male students wearing a suit can fasten their jacket so there is less temptation to keep adjusting clothes. It is helpful to empty trouser pockets so that there are no keys or loose change to jangle.
- Women are advised to tie back long hair so there are no loose strands to pull on, and to avoid necklaces – hands held up to the neck twisting beads are a clear demonstration of nervousness.

Nervous presenters have a tendency to wriggle, shifting from one foot to the other or leaning on furniture. It is possible to avoid this by standing with feet shoulder-width apart, keeping both the ball of the foot *and the heel* firmly on the ground. If you can do so, stand far enough away from furniture so that you cannot lean or swing on it.

The more nervous you feel, the more important it is to take every possible opportunity to practise your presentation skills. This is especially true of situations early in your MBA where work is not assessed. The authors have noticed that nervous students tend to avoid presenting early on in their MBA programme, thus making it even harder when they are obliged to present under exam conditions towards the end of the course. Our suggestion – hard though some may find this – is to take up every possible opportunity to present in class, at work or to other students. In this way, you will gradually build up your ability and confidence in presenting so that you improve your skills and confidence as you go along.

11.6 The confident presenter who obtains

low marks

In every cohort of MBA students there are several who are good at making presentations in other situations but gain low marks in the context of their MBA. This group of students expect a good result because they are confident public speakers and are often shocked when they receive low marks. The main reason for the poor grades awarded to this group is lack of practice and preparation. Well used to speaking at work, or giving witty after-dinner speeches, where the amount of time allocated is flexible, confident presenters will formulate the gist of what they want to say, without undertaking any formal rehearsal because that is usually sufficient in other contexts. However, for an MBA presentation a good general grasp of the subject is not enough to guarantee success, because your examiners can only give you credit for what you tell them and you have only a limited opportunity to make your points. Thus, confident speakers who are underprepared lose marks because

- they have not allocated their time precisely enough;
- they spend too much time on issues that are less important (such as the background and context of the problem) and not enough time on the key issues (their methodology and use of academic literature);
- they have not thought through carefully enough beforehand what they should include and what should be left out;

- they are on less familiar ground than at work, so they cannot rely on previous experience of the subject to 'get them through'.

If you are someone who is a good public speaker, but are failing to get the marks you had hoped for, consider carefully whether the above points apply to you. If so, it is worthwhile following the advice given above and setting out your presentation carefully and precisely so you are sure you have given time and space for the points you wish to make, then rehearsing it out loud in front of a friend or colleague so that you are well prepared on the day. This should make a significant difference to your MBA grades.

11.7 Dealing with questions

At the end of your presentation, you will be required to answer questions on your work put to you by your panel. This is often the part of the presentation that students find most difficult as academics can seem very challenging, and it is worth trying to explain the reasons for this below.

Academics are used to defending their work in public, because that is part of their job. They are used to speaking in front of large audiences – not just to students but at international conferences. Academics expect, at these events, to be challenged about what they have said. This means that they become practised at fielding questions and criticisms relating to their presentations in a public setting. As a result, they sometimes forget how painful this can be when it is experienced for the first time, so they may seem tactless when dealing with student presentations.

One way of dealing with this is to try and anticipate what questions you might be asked and think about how you might respond. We would not advocate preparing word-for-word answers (because it can be off-putting for you if the questions are not quite what you had hoped) but would suggest 'thinking around' the relevant areas. If you are uncertain what these might be, get your colleagues or fellow students to come up with some ideas and have a go at answering them when you do your run through. Just the act of trying to answer questions and defend your work to others will improve your ability to do this in a general sense and will prepare you for your presentation day. Above all do ensure that you do not allow yourself to become over-defensive when challenged by your panel. Remember that their challenges are not personal and that they will see a robust line of questioning as a 'normal' part of their job. It is therefore very important that you remain calm and pleasant, defending your point firmly but not in a defensive manner.

11.8 Summary

This chapter has provided general guidance about how to improve your presentation skills, especially if you are a nervous presenter, or if you are usually a confident speaker but are receiving unexpectedly low grades in your MBA programme. The key points to remember are as follows.

- Always prepare thoroughly and well in advance so that you are clear what to include and what to leave out.
- Practise your presentation out loud and to time beforehand.
- Be prepared to defend your case without adopting a defensive manner.

How to pass your exams

'I made sure I was very familiar with the work of several key authors such as Michael Porter. This was really useful at exam time as I was able to apply their thinking to the questions and add in a few quotes which probably pushed my grades up.'

Andrew, University of Central Lancashire

Introduction

For most students, no matter how clever they are, the greatest worries about succeeding on MBA programmes are focused on the exams. For readers who gained an undergraduate degree only a few years ago, exams may be a recent memory. For others, in particular those studying part-time, it may have been ten years or more since they felt the prickle of fear on entering an examination room.

Whatever the length of time since *your* last exam, you will probably remember the feelings of nervous tension that accompanied it. Be reassured that you are not alone. Most of your MBA colleagues will be just as worried as you are, and even A-grade students can be reduced to a quivering jelly at exam time. To be fair, these anxieties are often founded in a realistic concern that exam grades do not always provide an accurate reflection of ability. At Lancaster, we have observed that, on average, a student's 'typical' grade will drop by around 10% in an exam situation, i.e. students who would usually be expected to attain around 70% in assessed coursework would probably receive only 60% under exam conditions. This is not necessarily a problem for someone who drops from 65% to 55%, but it could cause real difficulties if you are a borderline student, or hoping for a distinction grade. Often, however, this drop is needless, caused by poor exam technique or inappropriate preparation.

Purpose of this chapter

The purpose of this chapter is to help you re-position the concept of 'exam' in your thinking – from an event over which you have little control, and in which 'luck' plays a significant part, to a process which you can manage successfully. In Chapter 12 we provide you with the following.

- A series of straightforward guidelines for optimizing exam performance, including 10 Golden Rules for exam success.
- Handy tips about preparation for exams and dealing with their aftermath.
- An overview of what examiners are looking for when they mark your paper and how to meet these expectations.
- A special section for students who find exams particularly difficult.

These simple strategies may sound very obvious when you read them, but it is remarkable how often students fail to consider them on the day, thus lowering their grades substantially. The guidelines are easy to follow and do not need to involve you in too much extra work. Following them will help you to overcome nerves and improve your examination grades.

The chapter closes with some detailed advice for those who wish to raise their grades by more than 10%. This will assist students who fear they may be 'borderline' cases, or those who usually do well in their written and classroom work, but are concerned that they may be unable to perform to high standards in the exam room.

12.1 Passing exams – good luck or good management?

The perceived difficulties of taking exams are usually twofold. In the first place, your memory is being tested. Everyone feels scared that on the big day they might forget something really important – or worse, everything they ever knew. In the second place, candidates will be placed in a room under unusual conditions and asked to write for three or four hours, a length of time which can appear dauntingly long before the exam begins, but far too short as time runs out, especially if all the questions have not been answered fully. Such fears are exacerbated by the fact that exams are commonly described in terms of the wheel of fortune, as an event over which the candidate has little control, like throwing dice. These beliefs are reinforced by well-meaning friends and relatives, who send good luck cards and mascots (which are sometimes even taken into the exam room by candidates, as if they might make a difference to the result). The feeling that, as an examinee, you are at the mercy of fate is probably intensified by the fact that the exam-marking process is shrouded in mystery. Once you have handed your paper in, you are unlikely ever to see it again. You will be given a grade, but this is rarely accompanied by any written feedback, so it is difficult to analyse how and where you did well, or badly, and why.

Although there may be an element of luck attached to your exam (getting one question you had hoped for, for example), this should not substantially influence the exam experience, or the outcome. This is because the ability to pass an exam, like so many other things in life, is based not on luck, but on skill and confidence. Like driving a car, or hitting a ball accurately, some students may acquire this technique more easily than others, but just about everyone, if they set aside some time to practise, will be able to improve their performance.

There are probably two common ways in which MBA students are expected to take exams. The first is the traditional method of revising, going alone into the exam room and answering unseen questions on a given subject. The second is a more recent approach whereby initially, students work in groups on a case study, about which they are then examined individually. Some case study examinations *may* allow for an 'open book' approach (where students are permitted to take notes and references into the examination room with them). In either eventuality, the exam itself will be pretty similar – you will have three or four hours in which to present yourself in the best possible light to the examiner, and whether or not you pass or fail your MBA could depend on it.

Tips for optimizing examination performance

We will offer you some tips on the following areas to improve your exam performance:

- exam preparation
- on the day of the exam

and some strategies for coping after the exam.

Exam preparation

Tip 1 Begin your revision in good time and try not to leave it all until the last minute. It is worth noting here that revision should begin at least three to four weeks before the date of the exam and that you should resist the temptation to 'question spot'. Allow yourself sufficient time to revise broadly, familiarizing yourself with key themes, frameworks and authors as you go along. Even if it is an 'open book' exam, it is still advisable to review the main theories, authors, models and ideas you may need to draw upon, enabling you to understand key debates surrounding these areas.

Tip 2 If it is an 'open book' exam, resist the temptation to take your complete library into the room with you. You will not have time to read up on things during the exam. Instead, carefully select a small number of essential texts to take with you on the day and mark up the key models so that you can find them quickly.

Tip 3 If your exam (like most) is not 'open book', try and learn one or two quotes from key authors which you can include, *where relevant*, in your exam paper. For example, in a section on marketing and strategy, you might include the following quote by Philip Kotler:

> '*It is more important to do what is strategically right than what is immediately profitable.*' Kotler (2000)

Tip 4 Seek out some old exam questions and practise answering questions to time. If you can't get hold of old exam papers, don't worry. Just use one of the 'question and answer' sections included at the end of chapters in one of the classic MBA texts such as Philip Kotler's *Marketing Management*. The purpose here is not simply to check out what you know about the subject (though this is one way of doing that!), but principally to get you used to organizing your thoughts on paper within a short timescale. Shut yourself away for an hour or 45 minutes, answer the question as effectively as you can and don't be tempted to stop until your time is up. Once you have finished your 'practice exam', you can refer to your revision texts and notes and check the following:

- Did you understand what the examiner was asking for and have you covered all the key points?
- Did you complete the question on time?

- Is your answer clear and legible? Did you underline headings and spell names correctly?
- Did you remember to reference the main topics, citing academic authors and frameworks where appropriate?
- What could you do differently to improve your performance another time?

Tip 5 Unless your faculty allows the use of keyboards in exams (and most institutions still forbid this), get into the habit of writing by hand a few weeks before the exam date. Hand writing for three or four hours comes as a shock to those of us used to working exclusively on a PC.

Tip 6 Where possible, team up with a course colleague or a small group and share revision sessions. If you can work with students who get similar grades to yourself, but have different interests, this would be ideal, as you can help one another out. For example, a student who is comfortable with 'hard' data (e.g. financial forecasting) but less happy about writing in essay form might find it beneficial to work with a colleague who lacks confidence about financial reporting and control, but enjoys developing arguments about 'softer' subjects such as leadership and organizational behaviour. If you are lucky enough to be preparing for an 'open book' exam, you will find it useful to create a portfolio of key frameworks, theories and references for each subject. This will act as an invaluable *aide-mémoire* on the day of the exam.

On the day of the exam – Ten Golden Rules for success

Rule 1 Make sure you are certain about the location of the exam venue.

This sounds obvious, but for part-time or distance-learning students arriving at a large and unfamiliar campus, it may be important to allow extra travel time to avoid needless anxiety on the day. One very able student, arriving at a sprawling city campus, missed the start of his exam because he had got into a muddle about the room number. By the time the student had realized his mistake and arrived at the correct venue, nearly an hour of the three-hour time allocation had been lost.

Rule 2 Calm yourself before you begin.

Once you arrive at your seat, spend a few moments composing yourself while you wait for the exam to begin. Don't look to see what others are doing. Make sure you can see the clock and synchronize your watch with it. Take a few deep breaths in and out if you are feeling very nervous.

Rule 3 Once you have been given permission to start, turn over your paper and read it through carefully from start to finish.

Check that there aren't any questions on the reverse of the exam sheet. Make sure that you are quite clear exactly which, and how many questions you are supposed to answer. If you answer more questions than you need to, you will not be given extra credit – and you will lose precious time. Ensure that you understand which questions are compulsory and where you are offered an element of choice. Check the weighting of marks allocated to each question.

Rule 4 Once you have read ALL the questions on the paper thoroughly, decide which ones you are going to attempt.

Go for the 'best' first and so on, in order of preference. It will improve your confidence if you get off to a good start. Notice whether any of the questions you intend to tackle contain more than one part. If so, be sure to attempt each section, because marks will have been allocated to each part and students can miss out on marks because they overlook this.

Rule 5 Read each question again carefully before you start writing.

Think hard about what the examiner is asking for and include only *relevant* information in your answer, not everything you know about the subject. Do not answer the question you hoped/planned for during your revision – answer the question that is on the paper.

Rule 6 Draw up a short plan before starting each question.

This will help you to discipline your thinking and to decide in advance what to include in your answer and (just as importantly), what to leave out. Remember to apply the *relevant* academic frameworks and to demonstrate, where possible, that you can associate the right author with the concepts/theories that you are using. Try to frame your arguments concisely and be clear about the point you are making. If you are using numbers or tables to illustrate your point, ensure that you refer to these within the text, so that the examiner understands what they are there for.

Rule 7 Apportion your time properly.
You need enough time to construct your plan, write your answer, draw conclusions and read through what you have written. Make sure that you attempt the number of questions you are asked to answer. Remember that you will get *no marks at all for questions that you do not attempt*, so be strict with your time allocation. If you are supposed to answer four questions, don't be tempted to 'overrun' on the first three. If that means you have to leave one question unfinished and move on, do so. You will get some credit for giving a partial answer, but nothing at all for sections that you have not started.

Rule 8 Make sure that your paper is presented as neatly and clearly as possible.
Although computers are starting to creep into the exam room, many business schools still insist on hand-written scripts. Take a ruler and a supply of pens with you. Ensure that you have written a *clear heading* at the start of each question, so that the examiner knows exactly which question you have attempted, and try to write neatly. This may sound obvious; however, we are writing as examiners who have struggled to read many pages of illegible exam scripts, facing our own initiative tests in terms of which unlabelled answer might refer to which question. We can verify that, in the heat of the moment, some candidates *do* forget to include this vital piece of information. This is important – not just so that the examiner can be sure of matching the right answer to the appropriate question, but because a badly set-out exam script gives the impression of a badly organized student who is unable to present their thoughts clearly.

Rule 9 Try to avoid looking at other candidates in the exam room.
If you are stuck, it can be disconcerting to see everyone else scribbling away, apparently with confidence. It can be even more disturbing to be aware of colleagues in distress, which does occasionally happen. If you do become aware that someone is upset, remember that this is *not* your responsibility, even if they are well

known to you. It is the job of the invigilator to take action if necessary – your task is to use the three- or four-hour exam time as productively as you can for your own benefit.

Rule 10 Leave 5–10 minutes at the end of your paper for reading through.
You should also check spelling and the legibility of your handwriting.

Tips for excellence!

In the first paragraph of your answer, state briefly how you plan to structure your essay and how this relates to the question. This will not only guide the examiner, but will serve to keep you on track. As you write, use sub-headings to help you structure what you are trying to say. In the final paragraph, summarize your answer and re-state how you have answered all parts of the question.

After the exam – strategies for coping

Once you have completed your exam, there is nothing more you can do to improve your performance in that subject. Therefore, unless the exam is a 'mock', or practice exam, resist the temptation to discuss your answers with other colleagues or tutors/faculty advisers. Doing a 'post mortem' on what you have written is a pointless exercise because it is usually upsetting. There will *always* be something you have 'missed out' but which someone else has included and worrying about this is wasted energy. Having handed your paper in, there is nothing you can do to change what you have said, so you may as well put it out of your mind and focus on what comes next. This is particularly important if you are taking a series of exams. If you are unable to hold off from joining in the post-exam discussions, don't take what others say to heart – go home and get on with revising for your next paper.

12.2 What the examiners are looking for

The old (and comforting) adage about examiners is that they are not seeking to uncover what students don't know but to discover how much they *do* know. This may provide solace for those who are worried they have not revised sufficiently, but it is not strictly accurate. Doing well in exams is not only a reflection of how much you can remember. As well as trying to elicit what you know, the examiner is interested in testing *how effectively you can analyse problems* and how well you can articulate your thoughts, when working to a tight timescale.

The truth is that examiners are also looking at *how well you can perform on the day*. While this may appear to be bad news – just increasing the pressure on you as an exam candidate – it need not be as alarming as it sounds. As we have already discussed, there is a great deal that even the most nervous examinee can do to improve exam performance, and for those who find it difficult to memorize large amounts of information, it should be heartening to realize that it is not how *much* you know on the day that counts, but how well you can *apply* it.

What examiners are seeking – especially in essay questions – is evidence of understanding and intellectual analysis. They will be looking at how clearly your answer is structured and at how well you have understood and interpreted the questions you have been set. At MBA level, there may not be one 'right' answer to your question. It is likely that you will be asked to produce a defensible argument about the topic under consideration, from which you can draw some sensible conclusions. You would be expected to have considered the relevance of pertinent management frameworks as you developed your discussions. This is a very different concept from simply showing that you know a lot about the subject (or that you didn't know very much, but have included anything you could possibly think of).

How to plan and structure answers to exam questions

To give you an example, we have assumed that your MBA programme has included a module on leadership and strategy (and that you have been given a pre-examination case study to read). We have set a 'mock' question in relation to this, and given an outline of a range of different answers and the kind of marks they would have been awarded.

Exam Question: *Discuss the leadership styles displayed by Steve Jobs (Apple) and Bill Gates (Microsoft) respectively. In relation to the sustaining of competitive advantage, whose approach do you think has been the most effective, and why?*

What the examiners are seeking here, is evidence that you have analysed the question in a logical manner and that you can frame a coherent argument, built around *relevant* theories of management, to support your view. The examiners are expecting you to structure a discussion around leadership and its relevance to success. An excellent answer would conclude with some general observations which could be applied to other situations. There are several approaches you could choose in response to this question, focusing on a wide range of literature. Below are some examples of how a range of examinees might have interpreted the question and planned their answers. An indication of what sort of grades they might have received is given below.

Fail
Panic. Barely read the question. Rush into the answer without stopping to think about anything. Do not attempt to structure your answer. *Either* engage in long drawn out discussion about Jobs and Gates, without reference to competitive advantage, *or* put down anything you can remember on competitive advantage (in any order) without reference to the management concepts you have learned on your MBA programme. Write quickly and untidily so that your exam script is difficult to read. Do not include a proper heading at the top of this section, so the examiner will be uncertain about which question your answer refers to. Do not allocate your time properly and either abandon this attempt because you need to move on to another question, or overrun so that you are short of time at the end.

The plan/strategy for a bare pass/fail
Rush into the question without stopping to think about what the examiner is looking for. Write down what you can remember about Gates and Jobs. Write down anything you can remember

about leadership. Write down anything you know about competitive advantage. Either draw no conclusions at all, or provide conclusions that do not relate to what has been written up to this point. Leave no time to read through at the end.

The plan/strategy for a good/satisfactory answer
Describe the leadership styles of Jobs and Gates. List some of the features of competitive advantage and explain how far the respective leadership styles of Jobs and Gates might achieve this. List some of the writers on competitive advantage such as Michael Porter. Explain whose style was most effective and why. Try to leave enough time to read through at the end.

The plan/strategy for a very good/good answer
The leadership of Jobs and Gates
Define the leadership styles of Jobs and Gates and relate these to some of the theories on leadership.

Competitive advantage
Bring in the relevant management frameworks relating to competitive advantage (e.g. Porter's 5-forces model). Explain how far the leadership styles of Jobs and Gates were likely to lead to sustainable competitive advantage (and where they might fail).

Conclusions
Draw some conclusions about whose style was the most effective and why.
Allow time to read through and label headings.

The plan/strategy for a distinction grade answer
Total time allocation: one hour (plan: 10 mins)

Jobs, Gates and leadership styles
Define and *analyse* (by comparing and contrasting) the leadership styles of Jobs and Gates. Frame this discussion around different styles of leadership 'in action' (give brief examples) and show evidence of broad reading by referring to the work of less conventional authors, such as Machiavelli, on leadership. (15 mins)

<u>Leadership styles and competitive advantage</u>

Consider the *consequences* of the respective styles of Jobs and Gates in relation to how successfully (or otherwise) each managed to achieve/sustain competitive position and why. Draw upon *relevant* management frameworks which will shed light on these discussions, e.g. competitive advantage: Michael Porter (5-forces model? –useful in that it relates to external environment); also bring in vision and goals: Waterman (7S theory?).

Frame a convincing argument as to whose style was/is most appropriate for the achievement of sustainable competitive advantage and *why*. Include some relevant quotes from key authors and reference this. (15 mins)

<u>Conclusions</u>

Draw some conclusions about the general lessons on leadership and competitive advantage which can be learned from this and could be applied to other situations. (10 mins)

Read through. (10 mins) Ensure that each word is legible and that each section is clearly labelled.

12.3 Exam techniques for students in difficulties

It will be evident from the above that the key to exam success lies not merely in learning about the subject under consideration, but in the ability to remain calm and to structure your argument with some intellectual rigour. Most readers will find that they can combine the tips for optimizing exam performance with an understanding of what the examiners are seeking and this will help their exam performance. For some students, however, the exam remains a daunting prospect, either because they are overcome with nerves, or because they desperately need to raise their grades. The final section of this chapter is written for those who need extra help with their exam technique and who wish to invest additional time in this. In contrast to most of the advice given in this book, the adoption of the technique described below can take up quite a bit of time (maybe several days in total), depending on how fully you decide to follow it.

Real-life example: Exam techniques for students in difficulties

Example 12.1

Beth was in her late thirties and doing her MBA part-time. A single mother with two children, she also held down a demanding job in the IT industry. Time was at a premium and Beth was already making big sacrifices in order to manage her MBA on top of everything else. Since her schooldays, although her coursework had always been good, Beth had never done well in exams and was usually a 'borderline' candidate. She was unsure why this was the case, but it made her increasingly nervous every time she had to face an exam. Given the huge cost involved in taking the MBA, which was funded by her company, she did not feel that she could afford to take the risk of failing the exams and sought the help of her MBA tutor/adviser, Luke. Luke gave Beth some advice which she followed (working with Caroline, one of the authors of this book). Although at first Beth found it hard to believe, the new techniques (described below) really worked. Eventually, Beth raised her grades from around 50% (the passmark) to 70% and over (a distinction grade).

Raising your exam grades – techniques for those who need extra help

Choose a partner

This technique works best if you first choose a study partner (or partners). Since you need to work in pairs, two or four is suggested as a good number. These must be people you trust and get along with, who have a similar interest in improving exam confidence and raising exam grades. If you are unable to work with someone else, perhaps you could still get an MBA colleague to comment on your attempts at exam answers (even though they may not wish to try the 'technique' for themselves). Failing this, try and critique your own work by putting it down when you have finished and evaluating it the following day through 'fresh eyes'.

The Golden Rules

The technique then involves familiarizing yourself with the Golden Rules and other guidelines for optimizing exam performance described above. It is true that the more nervous you are, the more likely it is that they will be forgotten once the exam has started. However, the techniques described below will help you to internalize them so that you put them into practice as a matter of course every time you attempt an exam question. Now that you have considered the Golden Rules, you and your study partner are going to learn how to apply them in practice.

A list of words and questions

To begin with, see if you can get hold of some old exam papers. Choose a question relating to a topic which you are both reasonably comfortable with, such as strategy, and talk through with your study partner what you think the question is asking you to do. Spend about 15–20 minutes on this. Don't try to write the answer at this stage, just jot down a few key words and questions that will help you clarify what you think the examiner is asking for. Supposing the question was worded as follows:

Example 12.1 Exam Question:

'Strategy optimizes competitive advantage and reduces the chance of failure.'

Discuss the above statement, using the relevant frameworks from your MBA course.

The key words here are *strategy* and *competitive advantage*. The examiners expect you to show that you know something about each of these areas and to give your view on how (or whether) they are linked. They have also given you broad hints about how they think the answer to this question should be presented. They have asked for a 'discussion', by which they mean they hope you will present some sort of argument giving both sides of the picture. They have also given a helpful reminder (which applies to any question on an

MBA exam) that a good exam answer will make explicit reference to the MBA literature that you have been taught on the course – so you should, at this stage, think back to what you have covered during your course and select the names of some writers on the subject areas, and the frameworks which you associate with them.

Your list of key words and questions at this point might look like this:

A list of key words and questions

Strategy ... competitive advantage...How are the two linked? Can you have one without the other?

Should we begin by defining strategy and competitive advantage?

Examples from MBA literature: Henry Mintzberg – deliberate and emergent strategy...Michael Porter 5-forces model...Robert Waterman 7S theory

Discuss – argue both sides of the case – list for and against

What are we considering? Does strategy affect competitive advantage and if so, how?

Do I need to decide whether you need a strategy to achieve competitive advantage??

Draw up a plan

Once you have completed your list, begin with your partner, to draw up a plan which sets out clearly how you might answer this question. Consider whether you have included everything you need to (and conversely whether you have included anything which is irrelevant). Take as long as you think you need to get this right. For the sort of thing you might include in the plan, you could refer to the examples given above, under the section on how to structure and plan answers to exam questions (page 205–208). We suggest that you spend up to 30 minutes on this, thinking your plan through very carefully.

Draft out an answer and compare notes

Once you have agreed your plan, sit down separately. Give yourselves one hour to draft out a complete answer to the exam question you have chosen. Do not be tempted to speak to one another, to break off from what you are doing, or refer to your revision notes (other than the plan and list of suggestions). At the end of one hour, sit down together and swap papers. Read through them carefully and then discuss jointly where each of you could have improved your answers. This conversation should focus on how well you have set out your answers, whether your arguments flow clearly and so on, as well as considering what you have included and what you have left out. Remember the Golden Rules.

Repeat the process

Choose a new question and start again, perhaps this time in a subject which is worrying you. Repeat the whole process as often as you feel you need to until you begin to feel more comfortable about interpreting exam questions and writing out answers 'to time'. After the first one or two attempts you should each begin to observe improvements in what the other has written, as well as feeling more confident about the whole process of producing a written exam answer. You should also be able to drop the preliminary process of including a list of key words and questions in your written work – these can be included in the Plan.

Try it on your own

Choose one or two new questions that you would both like to answer and attempt these by yourselves at home. Agree between yourselves how much time you will allow for the whole process. Work towards completing the question from start to finish in an hour or less, which is in keeping with the time you are likely to be allowed in a 'real' exam situation. Be strict with yourselves about timing (set an alarm clock, perhaps) and keep repeating this exercise until you are fairly comfortable that you can interpret exam questions, as well as planning your answer under exam conditions, within a given time. Once again, swap papers to see how well you have done and where there is still room for improvement.

Approach your tutor/adviser to see how you are doing

It is likely that you will receive more effective advice from your tutor at this stage (when you have done most of the hard work yourself!) than at an earlier stage. It is reasonable to give her/him one of your sample question/answers and ask for written or verbal feedback on your performance – though not necessarily to expect a 'grade' (this is a bonus if you get it).

By the end of this exercise you should be feeling a good deal more confident about the 'real thing'. You have had recent and regular practice at answering exam questions 'to time' and proved to yourself (and at least one other person!) that you *can* manage the exam process successfully.

12.4 Summary

In this chapter, we have taken a detailed look at how to manage the MBA exam system, reconstructing the concept of 'the exam' from an event entirely dependent on good or bad 'luck' to a process in which the exam candidate can take control and improve the outcome. A set of Golden Rules for exam success has been provided, as well as guidelines for managing exam preparation and the aftermath. Some insight into the mind of the examiner has been given, along with some suggestions for interpreting and structuring exam questions. Finally, a special technique has been explained, which involves extra work on the part of the exam candidate, but which is helpful for those who are seriously worried about exams and who wish to invest the time involved in tackling these worries.

How to manage and write your dissertation

13.7 Consultancy-based dissertations

13.8 The library-based dissertation

13.9 Summary

13.10 Conclusions

'The MBA opened my eyes to a huge wealth of concepts and tools that I had not been aware of. This has been of huge value in itself and has been the catalyst for me giving a much greater focus to ongoing learning. The dissertation introduced me to some key work on knowledge management that has become a central part of my work in IT management.'

Colin, University of Bradford, School of Management

Introduction

The writing of your dissertation is the culmination of all you have learned on your MBA. It provides you with the chance to draw together your learning and show the examiners what you know. The dissertation can make a significant difference to the overall marks you will get for your MBA, as it is often weighted heavily in comparison with other parts of the programme. For all of these reasons, the dissertation can seem less of an opportunity and more of an insurmountable problem, although after students have survived the experience they often tell us that doing the dissertation was the most fulfilling part of their MBA programme.

Purpose of this chapter

Through the advice offered in this chapter, we aim to help you to view the dissertation as a unique and exciting learning opportunity. Part of the key to being successful (and as a result, getting good marks for your dissertation) lies in the way you manage the process. Although beginning a big piece of work like a dissertation might seem daunting, it is worth remembering that you have already demonstrated your ability to write good assignments and have refined your study skills in order to be able to do this. This would be a good time to remind yourself of the basic requirements for an academic piece of work by reading the chapters on study skills and assignment writing (Chapters 2 and 10), in which tips for doing a literature search and review, choosing a methodology and analysing your findings are given.

Chapter 13 deals with more advanced questions and focuses on issues that are particularly pertinent to writing an MBA dissertation. The chapter begins with a general section giving advice which is relevant to all, and is followed by specific guidance relating to dissertations. Depending on whether you are a full-, part-time or distance-learning student, you are likely to be required to write one of three types of dissertation. These are:

- **A workplace dissertation** which is based on primary research undertaken by you in relation to a problem in your own workplace or similar setting agreed between yourself, your line-manager and your faculty adviser. This is likely to require you to deal with workplace politics.
- **A consultancy-based dissertation** which may be part of a work placement project, usually arranged on your behalf by your faculty adviser and a sponsoring organization, which you may be expected to tackle as part of a group. (*See also* Chapter 9.) This is likely to involve negotiating with the client in the sponsoring organization, other group members, and with your faculty.
- **A library-based dissertation** involving the collection and analysis of secondary data only, usually on a topic of your choice in agreement with your faculty adviser.

The issues involved in managing your dissertation may be different, depending on which type you are required to produce. For this reason, the three types of dissertation will be dealt with separately, although it is worth your while reading through all three sections, as some tips will be common to all.

The first two types of dissertation are similar in that they will both include some primary research (such as interviews or surveys), which will be designed, undertaken and analysed by you. This will then be linked to academic theory and written up in the form of a dissertation. The third type of dissertation, which is based on library research only, is dealt with separately since you will not be doing empirical research but will be relying on secondary sources.

This chapter therefore provides general guidance on how to manage your dissertation, including

- choice of subject
- evaluating ideas
- research design
- getting organized
- writing up
- tips on how to handle the three main types of dissertation.

13.1 Choosing a topic

If you are exploring an area of special interest, or have been given your topic as part of a work placement project, the subject of your dissertation might already be obvious to you. If not, you have the task of selecting one and, given how important the dissertation is to your final MBA grade, it is worth taking some time to think seriously about what interests you and what gives you the best chance to shine, before rushing to grasp the first topic that presents itself.

You could begin by setting aside some time (maybe with other students in the same position) to think imaginatively about what you might do. If you are feeling a bit short on imagination, you could draw on the suggestions below to get the 'creative juices' flowing.

Ideas for dissertation topics

- **Personal interests** Do you have an outside interest that would lend itself to management research? (*See* the discussion of disabled sports clubs below.)
- **The media** Is there a news story – either national, or local to you – that could provide you with a subject? For example, the impact of new gaming regulations on the economic prospects of a particular holiday resort.
- **Previous topics** Is there an issue from one of your previous lectures, or from your own assignments, that you have not fully explored and which could be extended?

- **Someone else's problem** Is there a company or an organization you are aware of that might welcome some help with management issues? A part-time student of one of the authors, who was unable to undertake research based in her own workplace, approached a small but expanding graphic design company and, using the relevant academic frameworks, developed a new marketing strategy for the firm.
- **A future career-oriented choice** Is there a topic you could explore, or an organization you could approach which will help you to develop specific knowledge or experience which will enhance your career prospects in the future?
- **An immediate career-oriented choice** Is there a possibility of finding a suitable project within an organization in which you would eventually like to work? Some MBA students, who have specifically targeted particular organizations, have succeeded in making the transition from student on a placement to employee.

Once you have created some ideas, you can evaluate these, choosing one or two which could be worked up into a proposal for discussion with your faculty adviser. As you evaluate the ideas, it is worth considering the following questions (which will probably be posed by your faculty adviser when you meet her or him, so may as well be borne in mind at this stage).

Evaluating ideas

- **Is my chosen subject relevant to my MBA?** Does my research topic meet the requirements of my programme? Can my choice be clearly and defensibly linked to academic issues that would be acceptable to MBA examiners, such as strategy or marketing?
- **Is my dissertation topic 'original?'** Examiners are not expecting MBA students to produce ground-breaking research in the manner of Einstein, but neither would they expect to see a repetition or description of work which has been done many times before. Perhaps you could think of a new approach which makes your research a bit different? Is there a marketing feature to something not usually regarded as marketing (for example, the

strategies utilized by environmentalist groups to *dissuade* people from consumerism)?

- **Is my chosen subject legitimate research?** Does it raise questions that are worth asking and will anyone other than me be interested in the answers? Ask yourself *'What* am I planning to do, and *why am I doing it?'*

- **Is my chosen subject feasible?** This question was discussed in Chapter 10 and is even more important to consider in relation to your dissertation. You need to think seriously about the practicalities of whether you can get access to the data you require. For example, you may wish to look at how effectively teacher-training colleges manage the school placements of student teachers. Before you begin, however, you will need to be sure that you have established good links with teacher-training colleges which would provide the necessary access to student teachers and schools.

- **Is my chosen subject ethical?** A good researcher will take the trouble to ensure that she or he has thought through the ethical dimensions of a research project. Could your research be harmful to participants? Have you asked permission from the relevant bodies? Can you promise confidentiality, where necessary? It would, for example, be considered unethical to speak to hospital patients about their treatment unless you had first gained permission from the relevant authorities and you could be sure of keeping the opinions and medical details of individual patients confidential.

Tips for excellence!

Is there a possibility of applying the learning from your dissertation in a wider sense? For example, a dissertation which looks at reducing waiting lists in one hospital might provide solutions for other hospitals – and could even be publishable. Emphasizing the wider relevance of the learning could add valuable marks to your dissertation grade.

13.2 Narrowing down your research brief and establishing your research question

Once you have narrowed down your choice of topic to one or two alternatives, it is worth writing a brief proposal which you can present to your adviser and, where applicable, to a potential client or host organization. Let them have a copy of this in advance of your meeting. By the time you meet, you may have done some reading around the subject, which will provide you with background knowledge. Your adviser will then be able to help you scope your research question so that you have a working title for your dissertation and one or two clear aims on which to focus.

At this point, some students might find it useful to visualize their idea by producing a 'rich picture'. The concept of using a visual representation of ideas in order to help clarify thought processes was developed by Peter Checkland, a Lancaster University professor who became an international expert in applying academic theory to 'real-world' problems, and this technique is summarized in his book *Soft Systems Methology in Action* (Checkland and Scholes, 1990). Checkland suggests representing important events, issues and relationships in a literal sense by sketching out by hand a drawing of the situation you are trying to explore, using arrows to link one set of issues to another. To show how rich pictures work in practice, we have provided an example below.

Example: Tim Brown's dissertation

Tim Brown was an MBA student with an interest in marketing and the not-for-profit sector. After his MBA, he hoped to gain a job fundraising for one of the major aid agencies. He therefore chose a topic which was feasible, attaching himself to a client who would benefit from the work he had chosen, which at the same time might be helpful to his career plans. Tim's brother Mark had lost the use of his legs following a motorbike accident. Mark used a wheelchair but was intent upon living an independent life and doing the same sort of things that Tim did. Mark had a job, and drove a car. He belonged to a disabled sports club named Supersport. Mark asked Tim to develop and implement a

---

marketing strategy to raise money for Supersport. He also urged Tim to produce a strategy that challenged the image of disabled people as helpless and dependent by presenting them as independent and capable. As he developed and evaluated this idea, Tim found it helpful to draw it as a 'rich picture', as shown in Figure 13.1.

Drawing the picture, and discussing it with his faculty adviser, assisted Tim in recognizing two problems. The first problem was that the request to change the image of disabled people presented a task which was not feasible within the scope of an MBA assignment, since it was too wide. In addition, Tim could not claim that the views of Mark and his friends at Supersport were representative of all disabled people. It was notable, for instance, that there were no visually-impaired members of the club. The second problem was that the members of Supersport were almost exclusively male. Thus, Tim did not have access to a female viewpoint and was unlikely to be able to change the public image of disabled women through his work with Supersport. In the event, he narrowed down his research brief so that it read as follows.

Tim Brown's dissertation – working title and research aims

Sponsorship for Supersport

- *Developing a marketing strategy to raise money for Supersport disabled sports club that is workable and cost-effective.*
- *Raising the profile of Supersport's members so that they can be seen as active, independent and valuable members of the community.*

In this way, Tim narrowed down the scope of his dissertation topic so that it was manageable and interesting, but still met the needs of his 'client', Supersport. Thus, the use of the 'rich picture' was beneficial, in that it helped Tim to visualize things clearly. This was particularly important for Tim who, because he was a part-time student, spent most of his study time by himself and did not have a 'learning set' with whom he could talk things through.

13.3 Research design

The advice given in this section does not attempt to cover all aspects of how to design your research, but it does provide some handy hints which could help you get started. Reading Chapter 9, on

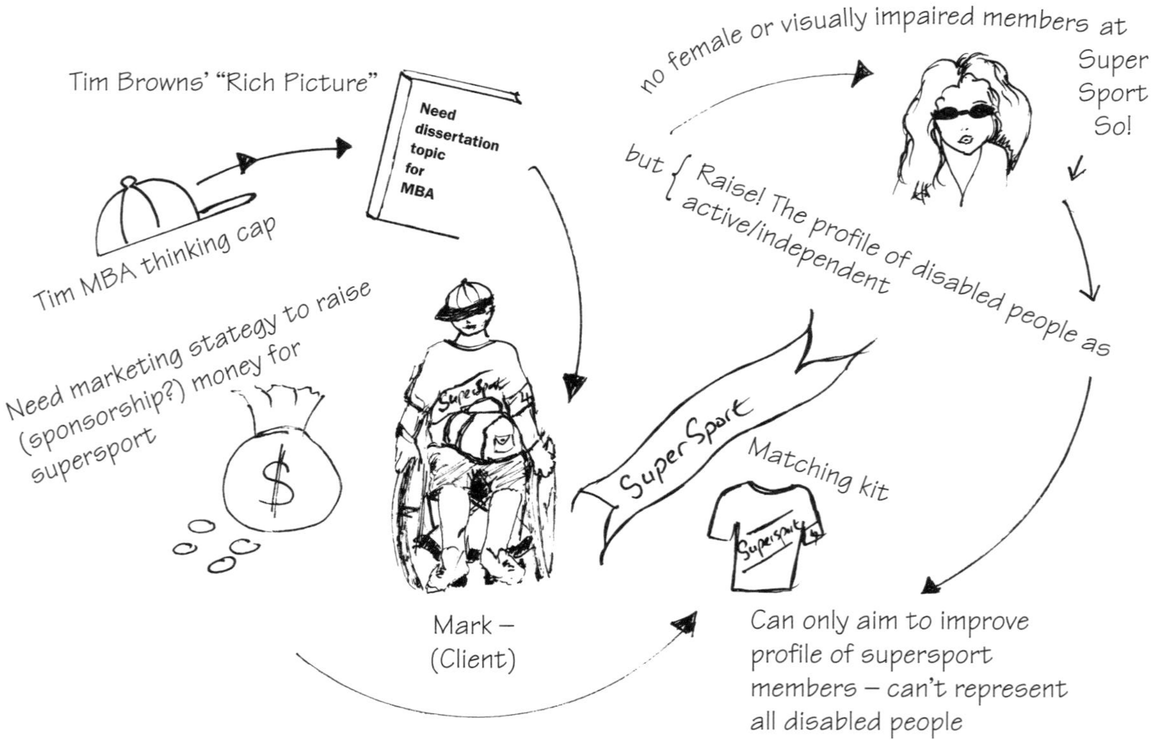

Figure 13.1 Tim's rich picture

Tim Browns' "Rich Picture"

Tim MBA thinking cap

Need dissertation topic for MBA

Need marketing strategy to raise money for (sponsorship?) supersport

Mark – (Client)

Super Sport

Matching kit

no female or visually impaired members at Super Sport So!

but { Raise! The profile of disabled people as active/independent

Can only aim to improve profile of supersport members – can't represent all disabled people applying theory to real-world problems, will help you relate your MBA learning to study in the field and give you hints about the range of research methodologies available to you. It is important to set aside time to think carefully about your research design. Hopefully, by the time you reach the dissertation stage of your degree, your faculty will have offered you some formal training on 'how to do' research. Some reliable texts on research design are recommended in the Further Reading section at the back of this book.

Research aims

If you are clear about the aims of your research, it will be easier to identify an appropriate research design. It is therefore worth spending time thinking seriously about your research question: *what* you are doing and *why* you are doing it, before you start to think about *how* you will do it. It has already been noted, in section 13.2, that one way of helping to define your research question is through drawing a 'rich picture'.

Once you are clear about your research aims, it is much easier to start to think about the approach you will take. Are you, for example, seeking to understand people's feelings and experience? Suppose you are looking at female users of executive hotel facilities. Do you want to know how it *feels* to be a woman staying in a hotel as an 'executive' guest? Or are you seeking to generate numerical data, such as how many hotels offer complementary spa treatments and whether these were available without prior booking? The first research question would indicate the need for *qualitative* research, in which you study the experiences of a small number of individuals in depth. The second suggests a *quantitative* approach, where you ask the same question to a large number of hoteliers, and/or female guests. Some research questions may indicate the need for several different approaches. This might be an option for MBA students working in groups, because each student can examine the research problem from a different angle.

Unless you can think of a means of scoping your research design so that it is manageable, however (as in Anita's example given below), this may be unrealistic for MBA students working alone, and a choice may have to be made about which research technique is most suitable.

Clarity about research aims will also assist you in choosing appropriate academic frameworks to help solve your problem – and avoid the temptation to try and 'fit' a much-loved theory around your research (*see also* Chapter 8 on case studies).

Whatever your research design, remember to write it up fully in the methodology section. As noted in Chapter 10, the examiner can only give you credit for what you tell him or her. Remember that academic examiners are interested in process – in *what* you did and *why* you did it – as much as in the end results.

Real-life example: Mixing quantitative and qualitative methods

Example 13.1

Anita is a practising manager who studied for her MBA part-time. A public sector executive, she is the State Director of Tourism in a major holiday resort. In her resort, Anita had already established a market niche for wheelchair users travelling with one able-bodied person. She wished to apply for grant funding to further improve facilities in her area. The aims of her dissertation were both market-led and socially entrepreneurial. Anita wished to attract money and increased visitor numbers to her town by enhancing its reputation as a resort attractive to wheelchair users and their companions. She also felt a social responsibility to make visits to her area as enjoyable as possible for this group of visitors. The main thrust of Anita's research was quantitative, as she undertook a detailed audit of the facilities available in her town, this enabling her to apply for grant monies, on behalf of hoteliers, to upgrade facilities. This was a demanding project and left little time for additional approaches. However, to complement the *quantitative* research, Anita wished to undertake *qualitative* research in order to establish how suitable the main tourist attractions in her area were for disabled visitors. She therefore borrowed a wheelchair for a weekend and spent two days using it, resisting the temptation to get out and walk when things became difficult. In this way, Anita was able to gain some understanding of what it *felt* like to be a user of the facilities on offer, as well as simply having an audit of what these were. Anita gained an excellent mark for her dissertation.

13.4 Getting your dissertation organized

It might seem as though you have a long time in which to complete your dissertation – but time passes very quickly when you are doing a big piece of work. As soon as you are sure about your research question and research design, it is essential to get started. Make appointments to see people well in advance and ensure that surveys are sent out with plenty of time to allow for late returns, analysis of data and writing up. Plan a research timetable and stick to it. Your faculty may be flexible about some aspects of your dissertation – but missing the deadline is unlikely to be one of them.

Managing the process of writing up

We have already included guidelines on writing up projects in Chapter 10 on assignment writing. These are relevant here and as an adjunct to these, it is worth noting that a long piece of work, such as a dissertation, is difficult to put together and, if you have worked hard on your research, you will have a lot of material to write up. The suggestions made here will be familiar if you have read the earlier chapters in this book, but they are worth summarizing again because without them you will struggle when it comes to pulling your dissertation together.

Reference as you go along

It is extremely difficult to sort out your reference section at the end of your dissertation when you are tired, and perhaps a little short on time.

Don't leave all your writing up until the end of your project

Start writing things down as early as possible, then the task of putting your dissertation on paper will seem less daunting. For many students, writing is the best way to start to think. The literature review is a good place to begin, as the theory you have read will help you to start to make sense of your research. Leave your introduction and conclusions until the end.

Don't leave all your writing up till the last minute

The amount of time needed to write up an MBA dissertation is rather like the amount of money needed on holiday – you will need more than you think! Failing to allow enough time to write up may mean that your grades do not reflect the amount of hard work you have invested in your research/project. This is very common and it is always clear to us when students have miscalculated the time needed to write up their work, as it is invariably their conclusions and recommendations that are weak, arguably two of the most important sections of the dissertation.

Keep in touch with your faculty advisers

This is especially important if you are stuck. They are there to help you.

Allow plenty of time to read through and edit.

Check that you are meeting faculty guidelines and check your grammar and spelling before you hand in your dissertation.

Tips for excellence!

Academic examiners set great store by neat and tidy references. Ensure that you are following the style required by your faculty and that all names are spelt correctly and dates and page numbers are included. Some examiners will flick through the reference section before beginning to read your dissertation, so it might even be the first thing they see. Create a good first impression and ensure that your reference section is perfect.

13.5 Studying alone

If you are a part-time, e-, or distance-learning student doing a research-based topic, the dissertation can seem a lonely process. Make sure you keep in regular touch with your adviser and, where possible, with other students. Although tackling different topics, they are likely to be experiencing the same problems and anxieties as you are. You will be able to offer mutual support to one another, as well as helping one another 'shape' ideas and thoughts.

13.6 Workplace dissertations

The following section will be useful to anyone undertaking a dissertation in a workplace setting. Particular attention is given to those who are attempting to produce a dissertation connected to their own employment. Some additional guidance is provided for those doing their dissertation as part of a short-term placement with an employer (section 13.7, below).

The main issue for practising managers to consider is the fact that their dissertation is connected with their paid employment, so the way it is managed could affect career status, as well as MBA grades. A further issue faced by students doing work-based dissertations is that of finding a topic which meets the requirements of both faculty advisers and line-managers, as well as being interesting to the student. In relation to the workplace, it is important to think carefully about the practicality of your topic in terms of its cost, scope and political acceptability inside the organization, and ethical responsibility. It is always a good idea to find a senior sponsor for your project inside the organization before you embark on it. If no sponsor is forthcoming, you need to ask yourself why this might be. Is it a political 'hot potato'? If so, it might be wise to steer well away from it and choose something less sensitive. Issues to bear in mind are described below.

- **Cost** – check out your budget before you begin. Perhaps you hope to undertake face-to-face interviews with colleagues from overseas? It may be that your company sponsor will find a budget to pay for travel. However, make sure that any expenditure incurred is agreed (preferably in writing) so that you are not left with unexpected responsibility for research bills.

- **Scope** – has your line manager understood the scope of your dissertation (and the faculty constraints you must adhere to)? No matter how interesting the topic, you must finish on time and within your word limit. Don't be persuaded to take on too many of your company's problems. In our experience, the tighter the scope of a dissertation, the more depth you will achieve, and the more effective your findings and recommendations are likely to be. If the issue seems too big to tackle, choose an 'angle' and don't be afraid to modify the proposal made by your line-manager or sponsor. For example, supposing you worked in a college of further education and were asked by the principal to consider the problem 'How could the college be more effectively managed?' You might suggest looking specifically at the more manageable question: 'Could departmental budgets be managed more effectively than at present and, if so, how?'

- **Political acceptability** – as noted in Chapter 10, it is crucially important that you do not choose a topic which could limit your career. This is discussed below in the section on common problems.

- **Ethical responsibility** – as pointed out in section 13.2 above, all researchers must try to behave ethically in the way they gather information. For your own sake, it is particularly important that you consider this if doing a dissertation in your own workplace. For example, does your line-manager understand that the survey you are undertaking will be anonymous? Will those filling it in understand that the line-manager might see it? Have you made any promises that you can't keep (such as no job losses, if you are undertaking a service review)?

- **Sponsorship** – finding a senior sponsor who can 'open doors' for you and ensure that you have access to the right people and documentation may not be easy, but it is worth your persistence in doing so at the beginning of the project. Don't be afraid to approach a senior manager, even if you don't know him or her very well. In our experience, senior managers can be very helpful and interested, and doing an MBA project for someone of high status in the organization can often be extremely career enhancing later.

Common problems faced by those doing a dissertation in their own workplace

There are several problems which are commonly faced by those writing a work-based dissertation. As noted above, these are mostly connected with the 'politics' of undertaking an academic piece of work in the workplace.

The unhelpful line-manager

Not all line-managers are enthusiastic about part-time MBA students who hope to research and write a dissertation at work. This may be for a variety of reasons. If your manager has never gained an MBA, for example, there may be a lack of understanding of what is involved in a Master's degree. Your manager could feel that you pose a threat – there may even be an element of jealousy. Some line-managers may consider that MBAs take up too much work time, others may worry that once you have your degree, you will want a pay rise or will leave the company. Whatever your plans for the future, it is important to get your manager to support you while you complete your MBA and there are various ways of doing this. The key to winning this battle is to be aware that there may be sensitivities and, therefore, to tread carefully. One suggestion is to agree on a research topic which interests you, but where the benefits to your line-manager are also clear to *both* of you. Another is to enlist the support of a high-level sponsor of the project in the organization, or to ask your faculty adviser for help – he or she might be willing to speak to your line-manager on your behalf to try and smooth things over.

Unhelpful colleagues

Sometimes, colleagues who know you well find it difficult to take your research seriously. For example, they cancel appointments with you when something more pressing turns up, or they do not bother to fill in forms, questionnaires, or diaries they have been given. The best way of dealing with this is to make it clear that you take your research seriously yourself and, if possible, that it has the backing of a senior manager in your organization. For example, make appointments to speak to people well in advance, then write formally (preferably by letter, as opposed to e-mail) to confirm dates and times. If necessary, ask your sponsor to sign the initial letter about the research. In our experience this can work wonders for your colleagues' cooperation. If possible, book a meeting room for interviews rather than holding them in offices where you may be interrupted. Taking this kind of trouble makes it clear that the research is important not only to *you* but also to the organization. Formalities are difficult to ignore and most people would be reluctant to cancel an appointment at the last minute once the date has been agreed in writing.

Colleagues may also seem unhelpful because they fear that you are encroaching on their territory, or that your research may somehow disadvantage them. For example, if you choose to examine whether departmental budgets could be managed more efficiently, colleagues might see this both as a personal criticism (implying that they manage their budgets badly) or a first step towards cutting spending. It is important, therefore, to be clear about the boundaries of your research and to be honest about how it might affect others, both with yourself and them. If you have found a sponsor, remind yourself that this is your project, and don't allow yourself to be used for purposes other than the agenda you have set out to meet.

13.7 Consultancy-based dissertations

Dissertations undertaken as work placement 'projects' are often linked with a work placement at the end of a full-time MBA programme. You may be working individually, but are more likely to be working with a group of students, and may have only limited choice about which organization you are placed with and what you are doing there (although schools will try to place students in areas which reflect their interests and abilities).

Work placements can involve an array of exciting possibilities, and sponsoring organisations come in all shapes and sizes, ranging from small businesses to not for profit agencies, to international manufacturers of conglomerates and charities. Your project might be based on number of topics: perhaps the marketing of new products, the improvement of operations management or personnel management. Whatever your placement turns out to be, and who ever you are working with, make the most of it. It provides a great opportunity for networking and for applying theory to real world problems. Since work placements are likely to involve working as part of a small team, this also gives you the chance to work with other students in a 'live' setting.

Notably, a major element of doing well in a dissertation undertaken as part of a work placement will be your negotiating skills. You are trying to fulfil several aims, which may be conflicting. These will include producing a project which fulfils the needs of your sponsoring organization, getting along well with other members of your MBA group and, most importantly, fulfilling the criteria set by your university with regard to your dissertation!

The sponsoring organization

As an individual or a group, you will be required to fulfil the needs and expectations of the sponsoring organization with whom you are placed, and who can be regarded as your 'client'. You will be required to undertake a particular project and use this as the basis of your dissertation. Sometimes, it becomes obvious that the project set up by the client is not appropriate for an MBA dissertation (or for you personally, as in the case of Mohammed described in example 13.2). This may be because the scope of the task is too wide, or because the issues have not been thought through properly. For example, one group of MBA students working for a large motor company were obliged to change the focus of their project through no fault of their own because it fell foul of union agreements.

If something goes seriously wrong with your project, it is crucial that you let your faculty adviser know of this as soon as possible so that they can help you sort this out. You may also have a named contact within your sponsoring organization who can offer advice and support to help you through difficulties. By thinking things through calmly, you can sometimes resolve these problems yourself. The MBA group working for the car manufacturer mentioned above developed an idea for another project, which was acceptable both to the sponsoring employer, the unions and their faculty. They completed this successfully and all gained good marks.

When undertaking your dissertation as part of a work placement, it is important to make sure that the detailed parameters of your project have been decided between yourselves and your sponsoring organization. It is likely that this element of your project will be down to you (though your faculty may offer advice). This would involve details such as agreeing the number of hours you will be present at the workplace and the amount of study time you have available, budgets for travel, phone calls and so on. It is important that you tackle these issues at the beginning and that you discuss matters with your sponsor if you wish to make changes, so as to avoid unnecessary misunderstandings later on. It is also important that you are diplomatic and ethical in your dealings with those you are working with on a day-to-day basis. This is partly for your own sake because you may need their help and cooperation in completing your dissertation. It is also important to remember that you are only in your placement for a short time, but this is their livelihood and they may have to live with the results of your project for a long time! If you have not already done so, it is worth reading the section above on the workplace dissertation, as some of the issues about political acceptability and unhelpful colleagues will be relevant to those doing consultancy-based dissertations.

Real-life example: Negotiating with your client and other group members

Example 13.2

Mohammed and his group were interested in finance. They were allocated a placement within the financial branch of a conglomerate. Mohammed was upset to discover that their job was to market a new finance scheme which aimed to persuade low-income consumers to borrow money, because his own cultural and religious background meant that he regarded debt as something to be avoided at all costs.

He was therefore uncomfortable with the concept of tempting people to borrow money that they might be unable to pay back. Mohammed had to negotiate first with his faculty adviser and his MBA group to explain his position and gain their support. His MBA colleagues helped Mohammed to persuade the sponsoring organization to change his topic. In the event it was agreed that Mohammed's MBA colleagues would continue to work on the finance scheme launch, but that Mohammed would be given a separate project related to insurance products.

The other group members

It may be that you have already read Chapter 3, which includes a section on working with groups. If not, you may find it helpful to do so at this point. The key to working well with your MBA group is to communicate and to try and ensure that tasks are shared out as evenly and fairly as possible. No matter how busy or involved you are in your project, it is important to ensure that you meet regularly as a group to discuss project developments. Since you will all have to produce your own piece of written work at the end of your field-work, you will need to agree who is doing what. It may be, for example, that you all undertake the project jointly, but consider it from different angles in writing up your dissertation. For example, in launching a new credit card, one MBA student could consider the marketing angle, another the economics and a third the operational issues of how the bank deals with management of credit limits and customer contact. If in doubt, check with your tutor what is permissible within the rules of your programme.

The faculty

Although it is important to try and please the client and to get on well with your MBA colleagues, it is equally important not to lose sight of the fact that your faculty will be marking your dissertation, which may be worth up to a third of your entire MBA degree. It is important to take advantage of all the academic supervision offered by your faculty and, while demonstrating your ability to undertake a project in a mature and independent manner, not to hesitate to contact your faculty adviser if problems become serious. When you write up your dissertation, ensure that you adhere strictly to faculty guidelines about layout, word limit and content – even if this means your having to write a separate report for your 'client'. You do not want to lose marks because you have failed to meet university requirements.

> ## Tips for excellence!
>
> Keep a balance – if you can manage to fulfil the needs of your client, and build good relationships with those you meet during your placement, at the same time as getting on well with your MBA research group and meeting faculty requirements, you will be well on your way to getting high marks for your dissertation.

13.8 The library-based dissertation

Some MBA courses offer students the opportunity to research a theoretical topic for their dissertation, rather than conducting an organizational project. You may even have a choice between an empirical and a theoretical study. A library-based dissertation can be fascinating, as it offers you an opportunity to immerse yourself in an area of literature and become an expert in your field – but it is not an easy option. Sometimes the research is conducted in groups, but written up separately; alternatively, these might be individual research projects. Three examples of such dissertations that we have encountered recently have been 'The Role of Reflection in Management'; 'Conceptions of Time in Studies of Organizational Change'; 'Organizational Change and Strategy in a Major Charity'. We will return to the second of these later in this section, to illustrate how a group of students tackled the topic.

Whatever topic you choose, it is important that you check that it meets the criteria laid down by your school. For example, that it is relevant to the study of organizations, business, or management; that it is strategic in its focus; and that it adds to management knowledge. It should not, therefore, simply be a recycling of old ideas. This is more than an essay and your tutors will be looking for you to find an original angle on the topic, either through a new slant on old ideas, a new approach to an old question, or even a new question altogether.

It is often helpful to start with a question rather than simply a topic, as a question will put you into an investigative mode of thinking. If you are not sure of what question to pose, find out what has been written before on the subject, and what questions remain controversial in this field. You will often find that in academic papers, the authors have written a section toward the end of their article suggesting topics for further research. If the article is recent, the gap identified might still be open for some original research, so seize the opportunity.

Example of a library-based dissertation

Let us take, for example, the second of the three examples cited above, entitled 'Conceptions of Time in Studies of Organizational Change'.

This topic arose by chance, because one of the authors in her own research was undergoing a study of the sociology of time, in order to apply this to an organizational case study she was researching. As she got deeper into the literature on the sociology of time, she came across an article which suggested convincingly that studies of the management of change in organizations have all but neglected the temporal dimension, and need to focus urgently on the topic of time in order to further our understanding of change. Since this research question was not within the scope of her own research, she suggested to a team of four MBA students that they might be interested in researching it, since it was so relevant to their MBA and they had each declared an interest in the management of change. Keen to do it, and excited by the prospect of making a real contribution to theory, they embarked on the research. However, the first question they posed was, of course, where should they start?

Where should you start?

Whatever theoretical domain you find yourself investigating, you will need to scope out what reading will be relevant (and what will be irrelevant), how widely and deeply you need to read, and when to stop and analyse, in order to start to work out your own angle on the topic or question and your supporting arguments. Often doing a mind map, as discussed in Chapter 2 on study skills, can be helpful in this planning stage, as it enables you to identify the themes and sub-themes you will need to cover, and the interconnections between them. Of course, this process is always iterative, since once you start to read you will inevitably begin to identify additional themes that will become important in the development of your thinking. You will also, of course, identify issues that are interesting but will not be relevant to your argument. We will call these 'red herrings', as they will lead you down a cul-de-sac, if you are not careful, and waste a great deal of valuable time. You must look out for these and be prepared to set these aside in favour of those studies that you believe will help you to answer your research question.

In the example cited above, the group recognized straight away that they would need to study the literature on the management of change, as well as on the sociology of time, in order to decide how one might inform the other. They would also need to investigate whether any other authors had already studied the connection between the two.

They started with the literature on the management of change and, by the time they met with their tutor again, they had identified a number of key areas in the literature that they felt they should study. The key areas are described below.

1 Literature which they labelled *prescriptive*, since it offered suggestions for managers about how change should be managed in organizations. They noted that this literature often seemed to be unsupported by empirical evidence, consequently they were worried about the validity of these sources.
2 Literature which *theorized* the management of change, often containing explanatory rather than prescriptive models. They were finding these articles and books useful, as these drew heavily on other sources, and the group was starting to identify which earlier studies were considered to be seminal to the study of organizational change, and which must therefore be read for their own study to be convincing.

3 Literature which *described empirical studies* of organizational change, often written in case study form and focusing on a single organization. These varied in the depth and approach taken by the researcher, but most were longitudinal and gave useful illustrative data on change in action. Some of these studies had become classics in the field and the students noted that these had been quite heavily cited in the theoretical literature discussed in 2 above. The students felt that the last two literature types could be useful to the study of organizational change, but were uncertain what to do about the prescriptive literature, which appeared to lack academic credibility, yet they knew was being widely used within organizations.

4 In addition, their initial explorations had led them into a number of other fields and they had to decide which of these might be useful to their study. These were causing the group most angst, as they were divided about which might be important for their study. The additional fields they identified when reading initially around the topic of organizational change included literature on managing culture, business process reengineering, total quality management, empowerment, organizational design, leadership, gender, cross-cultural management, emotion and organizational learning, but nothing on time!

This was the point when the students needed to reflect and consolidate. Acutely aware of the size and scope of the literature they had uncovered, and the fact that they had not yet even started to investigate the literature on time, they recognized the need to drop some aspects of the study. With their tutor's guidance, they returned to the research question, and asked which of these domains might help them to address it.

After much deliberation, they eventually decided that all three literatures on organizational change itself should be retained. Even though type 1 was not well theorized, they decided to study these prescriptive books and articles for evidence of a time dimension, as it was these books that were influencing organizational policy-making. Reluctantly, however, some of the other topics which seemed interesting to them had to be dropped. In the end, they decided to focus only on the supplementary fields of organizational learning (since there seemed to be a number of references to organizational memory in this literature – a topic relevant to 'time'), and managing culture, as this seemed to illustrate well the notion of time as a social construction. Even though the other topics had seemed interesting, they recognized the need to focus their study to meet the tight time deadlines under which they were working.

Eventually, the group divided the topics between them and covered a huge amount of data in a very short time. However, the next hurdle that they met was what to do with all this data.

Analysing the data and sense-making

The next time the group hit a low point was after they had collected all their data and were now ready to make sense of it. They approached their tutor for advice on how this might be done. The tutor's advice is given below. (We have added our advice to students conducting individual dissertations *in italics*.)

- First share your data so that each of you has a complete overview of the main themes you have uncovered.
  (*If you are working alone sort out your data so that you can see a complete overview of what you have discovered.*)
- Now write all these themes on a flipchart (*or large piece of paper*) and look for the connections across the literature on change and the literature on time.
- Next ask yourselves what questions are raised in your data and what insights you appear to have found.
- Now ask, are these insights well supported by evidence from a number of sources or are they challenged by other research?
- Look for evidence which appears to contradict your line of argument. Ask why this might be. Does it really negate your argument or is it case specific? Can you offer counter evidence to support your line?
- Refine your arguments so that you are now clear what you want to write.
- Decide how you will divide the work. Often a library-based dissertation involves joint research, but results in individually written dissertations focusing on different aspects of the project.

(If you are working alone, ensure the scope of your dissertation is tight enough to allow you to achieve the depth that your research justifies. Discuss this with your tutor and, if in doubt, tighten the scope even further or you will only skim the surface of too many topics.)

- Ensure that you are each clear about the boundaries of your *own* dissertation and how these interface with those of your colleagues.
- Plan the structure of your report.
- Go ahead and write it up, meeting regularly with your tutor and the group to test out how robust your thinking remains.
- View this as an iterative process and always be prepared to return to the library when you see gaps appearing in your data or argument.

13.9 Summary

This chapter has provided general advice on how to manage your dissertation, including choosing a topic, evaluating ideas and designing your research. Emphasis is placed on the importance of being well organized, both in the way you manage your dissertation and the writing-up process. We have tried to illustrate the different types of dissertation that you might be required to produce on your MBA programme. Whilst this list is not exhaustive, it should give you helpful pointers on the likely challenges you might encounter.

13.10 Conclusions

As you reach the end of your dissertation, you are likely to be nearing the end of your MBA degree. We hope that your journey has been a good one and that the advice in this book has helped you along the way.

When we ask our students how they feel at the end of their MBA course, most tell us that they see the world through different eyes and have gained a sense of inner confidence as a result of the experience. We hope that much of what you have learned, not only in the formal part of the programme, but also from other students, will be invaluable to you in your future career. The intention of this book was to demystify the secret of getting an MBA, help you to improve your performance and reach your full potential. If we have contributed towards helping you negotiate the difficult pathways of an MBA, and to avoid some of the pitfalls that we encountered ourselves as students, the book will have achieved its aim.

References and further reading

Brassington, F. and Pettitt, S. (2000) *Principles of Marketing*, Financial Times/Prentice Hall

Bryson, J.M. (1993) *Strategic Planning for Public and Non-Profit Organisations*, Pergamon Press.

Burgoyne, J., Pedler, M. and Boydell, T. (1994) *Towards the Learning Company*, McGraw-Hill.

Buzan, T. and Buzan, B. (1996) *The Mind Map Book: How to use radiant thinking to maximize your brain's untapped potential. Radiant Thinking: Major evolution in human thought*, Plume Books.

Checkland, P. and Scholes, J. (1990) *Soft Systems Methodology in Action*, Wiley.

De Wit, B. and Meyer, R. (1998) *Strategy, Process, Content, Context*, Thomson Business Press.

Hill, T. (1991) *Production/Operations Management*, Prentice Hall.

Kotler, P. (2000) *Marketing Management*, Prentice Hall International.

Machiavelli, N. (1995) *The Prince*, Cambridge University Press.

Maddock, S. (1999) *Challenging Women: Gender culture and organisation*, Sage.

Malik, M. (ed) 1994, *Setting Priorities in Health Care*, Wiley.

Mintzberg, H. and Waters, J.A. (1985) 'Of strategy, deliberate and emergent,' *Strategic Management Journal*, July/Sept: 257–2.

Phillips, E. and Pugh, D. (1994) *How to get a PhD: A handbook for students and their supervisors*, Open University Press.

Porter, M.E. (1979) 'How competitive forces shape strategy,' *Harvard Business Review*, vol. 57: 137–43.

Strauss, A. and Corbin, J. (1990) *Basics of Qualitative Research*, Sage.

Additional reading

Choosing your MBA

If you have not yet decided which MBA you intend to apply for, we recommend:

Golzen, G. and The Association of MBAs (2001) *The Official MBA Handbook 2001/2002: The Association of MBAs Guide to Business Schools*, Financial Times/Prentice Hall.

Research design

For help with your research design, we recommend the following:

Denzin, N. and Lincoln, Y. (eds) (1994) *Handbook of Qualitative Research*, Sage.

Easterby-Smith, M., Thorpe, R. and Lowe, A. (2002) *Management Research: An introduction* (2nd edn), Sage.

Flowerdew, R. and Martin, D. (1997) *Methods in Human Geography. A guide for students doing a research project*, Longman.

Gill, J. and Johnson, P. (1996) *Research Methods for Managers*, Paul Chapman.

Kumar, R. (1999) *Research Methodology: A step-by-step guide for beginners*, Sage.

Saunders, M., Lewis, P. and Thornhill, S. (2000) *Research Methods for Business Students*, Financial Times/Prentice Hall.

Strauss, A. and Corbin, J., (1990) *Basics of Qualitative Research*, Sage.

Exam revision

For useful revision texts prior to examinations, we recommend:

The FT Mastering Series, which comprises useful books on finance, strategy, information management and marketing, published by Financial Times/Prentice Hall.

Strategy

For good general books on strategy, we recommend:

De Wit, B. and Meyer, R. (1998) *Strategy, Process, Content, Context*, Thomson Business Press.

Johnson, G. and Scholes, K. (1999) *Exploring Corporate Strategy*, Prentice Hall.

Porter, M. (1985) *Competitive Advantage: Creating and sustaining superior performance*, Free Press.

Waterman, R.H., Peters, T.J. and Phillips, J.R. (1980) *Structure is not organization*, Business Horizons, June.

MBA and study skills

For another good, general text on MBA study, we recommend:

Cameron, S. (1997) *The MBA Handbook: Study skills for managers*, Financial Times/Pitman Publishing.

Glossary

Action learning Used in the context of the MBA, this usually refers to learning from projects which involve applying management theory to real-world problems or issues. Action learning seeks to achieve two objectives: firstly to solve a problem in the organization and make a change happen; secondly to learn from the experience of conducting the intervention. Usually Action Learning involves becoming a member of a *learning set* in order to discuss the outcomes of organizational interventions with others and learn from the experience.

Cold called When a professor/lecturer is teaching the entire MBA class and calls upon one member, without prior warning, to speak on a given subject. This is likely to be related to work prepared the previous evening, such as a case study, and the student concerned will be expected to analyse this for the rest of her or his classmates.

Empirical research Information gathered about a situation (human or material), problem or phenomenon involving the collection of data from primary sources.

Faculty adviser A term often used interchangeably with *tutor*.

Five forces The five forces model was developed by Michael Porter as a sophisticated tool for analysing the competitive environment. It would be appropriate for examining the issue of whether a new market is likely to be profitable for a firm. The five forces model helps companies consider what competition already exists; the extent of threats from new entrants to the market; the potential for other companies to produce cheaper substitutes; and the relative strengths of buyers and suppliers.

Learning set A concept taken from the field of *Action Learning*. A group of students set up either voluntarily or by the faculty for the purpose of collaborative learning. The function, composition and duration of this group is variable, its prime purpose being to share insights and problems, collaborate in research, discuss the application of classroom learning to practice, discuss and share ideas on assignments and projects, work together on case studies, and practise for exams.

Methodology Methodology is a term for describing the way in which you approach your research. This would include a description of the research design and how you conducted your research, the academic framework used, the human or material subjects of your research and the procedures you followed to collect and analyze your data (surveys, interviews, etc.).

Pedagogy/pedagogical approach This term/phrase has been co-opted by the academic world to mean different forms of teaching and learning. We use the word here in relation to the different approaches you might encounter, from instruction at one end of the continuum, to self-directed learning at the other.

PEST Stands for the political, environmental, social and technical aspects of a situation. This is a simple tool for analysing the competitive environment.

Primary research Conducting some first-hand empirical research involving the collection of new data.

Qualitative research Research which explores human behaviours and emotions which are difficult to quantify – values, experiences and feelings, for example.

Quantitative research Research which involves the collection and analysis of numerical data which enables the researcher to measure, or 'put a figure on' a given situation.

Secondary research Drawing on and applying the research of others, usually data found in published sources.

SWOT Acronym for **s**trengths, **w**eaknesses, **o**pportunities, **t**hreats. A much-loved management tool used for organizational diagnosis, often prior to strategic decision making.

Tutor The faculty member responsible for your guidance throughout the programme, or for a specific module. The role of the tutor may vary from school to school, but he/she is likely to play a number of roles, such as advising on and assessing academic work, facilitating *learning sets*, running *tutorial meetings*, providing pastoral advice to students, giving lectures, providing continuity across modules of the programme.

Tutorial group See *learning set*.

Value chain The value chain was developed by Michael Porter and is considered a useful tool for analysing how effectively the activities of a business add value to its products. The value chain consists of nine elements which can be considered individually in order to establish whether costs can be cut or efficiently enhanced by reducing non-profitable activities. Businesses are advised to compare their value chains with those of their competitors.

Index